MEMOIRS
of the
American Mathematical Society

Number 1046

The Shape of Congruence Lattices

Keith A. Kearnes
Emil W. Kiss

March 2013 • Volume 222 • Number 1046 (fifth of 5 numbers) • ISSN 0065-9266

American Mathematical Society
Providence, Rhode Island

Library of Congress Cataloging-in-Publication Data

Cataloging-in-Publication Data has been applied for by the AMS. See
http://www.loc.gov/publish/cip/.

Memoirs of the American Mathematical Society

This journal is devoted entirely to research in pure and applied mathematics.

Publisher Item Identifier. The Publisher Item Identifier (PII) appears as a footnote on the Abstract page of each article. This alphanumeric string of characters uniquely identifies each article and can be used for future cataloguing, searching, and electronic retrieval.

Subscription information. Beginning with the January 2010 issue, *Memoirs* is accessible from `www.ams.org/journals`. The 2012 subscription begins with volume 215 and consists of six mailings, each containing one or more numbers. Subscription prices are as follows: for paper delivery, US$772 list, US$617.60 institutional member; for electronic delivery, US$679 list, US$543.20 institutional member. Upon request, subscribers to paper delivery of this journal are also entitled to receive electronic delivery. If ordering the paper version, subscribers outside the United States and India must pay a postage surcharge of US$69; subscribers in India must pay a postage surcharge of US$95. Expedited delivery to destinations in North America US$61; elsewhere US$167. Subscription renewals are subject to late fees. See `www.ams.org/help-faq` for more journal subscription information. Each number may be ordered separately; *please specify number* when ordering an individual number.

Back number information. For back issues see `www.ams.org/bookstore`.

Subscriptions and orders should be addressed to the American Mathematical Society, P. O. Box 845904, Boston, MA 02284-5904 USA. *All orders must be accompanied by payment.* Other correspondence should be addressed to 201 Charles Street, Providence, RI 02904-2294 USA.

Memoirs of the American Mathematical Society (ISSN 0065-9266) is published bimonthly (each volume consisting usually of more than one number) by the American Mathematical Society at 201 Charles Street, Providence, RI 02904-2294 USA. Periodicals postage paid at Providence, RI. Postmaster: Send address changes to Memoirs, American Mathematical Society, 201 Charles Street, Providence, RI 02904-2294 USA.

This publication is indexed in *Mathematical Reviews*®, *Zentralblatt MATH*, *Science Citation Index*®, *Science Citation Index*™*-Expanded*, *ISI Alerting Services*℠, *SciSearch*®, *Research Alert*®, *CompuMath Citation Index*®, *Current Contents*®*/Physical, Chemical & Earth Sciences.* This publication is archived in
Portico and *CLOCKSS*.
Printed in the United States of America.

∞ The paper used in this book is acid-free and falls within the guidelines
established to ensure permanence and durability.
Visit the AMS home page at `http://www.ams.org/`

10 9 8 7 6 5 4 3 2 1 18 17 16 15 14 13

Dedicated to Bjarni Jónsson,
For his work on congruence varieties,

To Ralph McKenzie,
For his work on commutator theories,

And to Walter Taylor,
For his work on Maltsev conditions.

Contents

Abstract

We develop the theories of the strong commutator, the rectangular commutator, the strong rectangular commutator, as well as a solvability theory for the nonmodular TC commutator. These theories are used to show that each of the following sets of statements are equivalent for a variety $\mathcal{V}$ of algebras.

(I) (a) $\mathcal{V}$ satisfies a nontrivial congruence identity.
(b) $\mathcal{V}$ satisfies an idempotent Maltsev condition that fails in the variety of semilattices.
(c) The rectangular commutator is trivial throughout $\mathcal{V}$.

(II) (a) $\mathcal{V}$ satisfies a nontrivial meet continuous congruence identity.
(b) $\mathcal{V}$ satisfies an idempotent Maltsev condition that fails in the variety of sets.
(c) The strong commutator is trivial throughout $\mathcal{V}$.
(d) The strong rectangular commutator is trivial throughout $\mathcal{V}$.

(III) (a) $\mathcal{V}$ is congruence semidistributive.
(b) $\mathcal{V}$ satisfies an idempotent Maltsev condition that fails in the variety of semilattices and in any nontrivial variety of modules.
(c) The rectangular and TC commutators are both trivial throughout $\mathcal{V}$.

We prove that a residually small variety that satisfies a congruence identity is congruence modular.

Received by the editor June 27, 2007, and, in revised form, November 28, 2011.
Article electronically published on September 18, 2012; S 0065-9266(2012)00667-8.
2010 *Mathematics Subject Classification.* Primary 08B05; Secondary 08B10.
Key words and phrases. Abelian, almost congruence distributivity, commutator theory, compatible semilattice operation, congruence identity, congruence modularity, 8congruence semidistributivity, Maltsev condition, meet continuous lattice, rectangulation, residual smallness, solvable, tame congruence theory, term condition, variety, weak difference term.
The first author was supported in part by NSF Grant #9802922.
The second author was supported in part by OTKA Grants #T043671 and #T043034.
Affiliations at time of publication: Keith A. Kearnes, Department of Mathematics, University of Colorado, Boulder, Colorado 80309-0395, email: kearnes@euclid.colorado.edu; and Emil W. Kiss, Loránd Eötvös University, Department of Algebra and Number Theory, 1117 Budapest, Pázmány Péter sétány 1/c, Hungary, email: ewkiss@cs.elte.hu

CHAPTER 1

Introduction

This monograph is concerned with the relationships between Maltsev conditions, commutator theories and the shapes of congruence lattices in varieties of algebras.

1.1. Shapes of Congruence Lattices

Carl F. Gauss, in [**23**], introduced the notation

$$a \equiv b \pmod{m}, \tag{1.1}$$

which is read as "a is congruent to b modulo m", to mean that the integers a and b have the same remainder upon division by the integer modulus m, equivalently that $a - b \in m\mathbb{Z}$. As the notation suggests, congruence modulo m is an equivalence relation on $\mathbb{Z}$. It develops that congruence modulo m is compatible with the ring operations of $\mathbb{Z}$, and that the only equivalence relations on $\mathbb{Z}$ that are compatible with the ring operations are congruences modulo m for $m \in \mathbb{Z}$.

Richard Dedekind conceived of a more general notion of "integer", which nowadays we call an ideal in a number ring. Dedekind extended the notation (1.1) to

$$a \equiv b \pmod{\mu} \tag{1.2}$$

where $a, b \in \mathbb{C}$ and $\mu \subseteq \mathbb{C}$; (1.2) is defined to hold if $a - b \in \mu$. Dedekind called a subset $\mu \subseteq \mathbb{C}$ a **module** if it could serve as the modulus of a congruence, i.e., if this relation of congruence modulo μ is an equivalence relation on $\mathbb{C}$. This happens precisely when μ is closed under subtraction. For Dedekind, therefore, a "module" was an additive subgroup of $\mathbb{C}$.

The set of Dedekind's modules is closed under the operations of intersection and sum. These two operations make the set of modules into a lattice. Dedekind proposed and investigated the problem of determining the identities of this lattice (the "laws of congruence arithmetic"). In 1900, in [**13**], he published the discovery that if $\alpha, \beta, \gamma \subseteq \mathbb{C}$ are modules, then

$$\alpha \cap \big(\beta + (\alpha \cap \gamma)\big) = (\alpha \cap \beta) + (\alpha \cap \gamma). \tag{1.3}$$

This 3-variable law of the lattice of modules is now called the **modular law**. Dedekind went on to prove that any equational law of congruence arithmetic that can be expressed with at most 3 variables is a consequence of the modular law and the laws valid in all lattices.

Dedekind did not write the law in the form (1.3), which is an identity, but rather as a quasi-identity: for all modules $\alpha, \beta, \gamma \subseteq \mathbb{C}$

$$\alpha \supseteq \gamma \longrightarrow \alpha \cap (\beta + \gamma) = (\alpha \cap \beta) + \gamma.^{1} \tag{1.4}$$

[1] In fact, Dedekind used the symbols $+$ and $-$ instead of $+$ and $\cap$.

Dedekind also discovered a useful "omitting sublattices" version of the modular law. It is the assertion that there do not exist modules $\alpha, \beta, \gamma \subseteq \mathbb{C}$ which generate a sublattice isomorphic to $\mathbf{N}_5$ (Theorem I.7.12 of [**1**]).

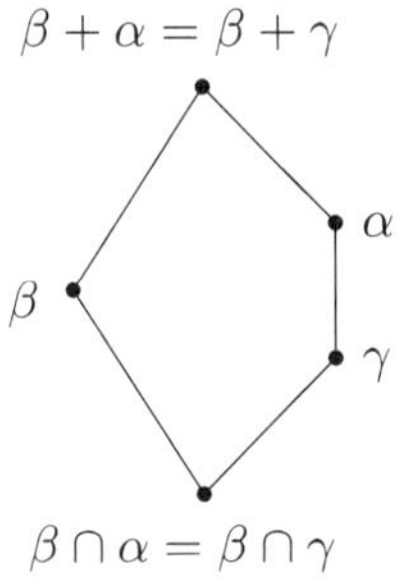

FIGURE 1.1. The lattice $\mathbf{N}_5$

More generally, a **congruence** on an arbitrary algebra $\mathbf{A}$ is an equivalence relation on the universe of $\mathbf{A}$ that is compatible with the operations of $\mathbf{A}$. Equivalently, it is the kernel of a homomorphism with domain $\mathbf{A}$. The set of all congruences is a sublattice of the lattice of equivalence relations on A. A **congruence identity** of $\mathbf{A}$ is an identity that holds in the lattice $\mathbf{Con}(\mathbf{A})$ of all congruences of $\mathbf{A}$. Dedekind's observation is that the modular law is a congruence identity satisfied by $\mathbb{C}$ as an abelian group. In fact, it is a congruence identity of any group, ring, vector space, Boolean algebra or lattice.

Dedekind's result initiated many lines of research in the 20th century, but to avoid losing focus we mention only a few. Garrett Birkhoff observed that the congruences of any group **permute**, meaning that $\alpha \circ \beta = \beta \circ \alpha$, and that any lattice of permuting equivalence relations is modular. This "explains" Dedekind's result to some degree, but in [**22**] N. Funayama and T. Nakayama showed that the lattice of congruences of a lattice satisfies the **distributive law**: for all α, β, γ

$$\alpha \cap (\beta + \gamma) = (\alpha \cap \beta) + (\alpha \cap \gamma), \tag{1.5}$$

which is stronger than the modular law, yet congruences of lattices need not permute. Thus permutability implies modularity, but not vice versa. Bjarni Jónsson then refined the results of Birkhoff and Dedekind by showing in [**37**] that any lattice of permuting equivalence relations satisfies the Arguesian law, which is stronger than the modular law. Conversely, he showed that any complemented lattice satisfying the Arguesian law is embeddable into the lattice of congruences of an abelian group. Here the **Arguesian law** is the 6-variable law asserting that for all α_i, β_i, $i = 0, 1, 2$,

$$(\alpha_0 + \beta_0) \cap (\alpha_1 + \beta_1) \cap (\alpha_2 + \beta_2) \leq \big(\alpha_0 \cap (\gamma + \alpha_1)\big) + \big(\beta_0 \cap (\gamma + \beta_1)\big) \tag{1.6}$$

where $\gamma = (\alpha_0 + \alpha_1) \cap (\beta_0 + \beta_1) \cap (\delta_0 + \delta_1)$, $\delta_0 = (\alpha_0 + \alpha_2) \cap (\beta_0 + \beta_2)$ and $\delta_1 = (\alpha_1 + \alpha_2) \cap (\beta_1 + \beta_2)$. Thus Jónsson solved the special instance of Dedekind's problem (of determining the laws of modules) which concerns only complemented lattices of modules. The full problem is still open, although in the same paper Jónsson gave an example of a finite noncomplemented Arguesian lattice that is not embeddable in the congruence lattice of a group, and in [**30**] Mark Haiman proved that lattices of permuting equivalence relations satisfy laws stronger than the Arguesian law.

1.2. Maltsev Conditions

Let $\mathcal{R}$ be the variety of rings and $\mathcal{A}$ be the variety of abelian groups. The fact that rings have a term-definable underlying abelian group structure may be denoted $\mathcal{A} \leq \mathcal{R}$, i.e., rings have the structure of abelian groups and more. A. I. Maltsev proved in [**63**] that a variety $\mathcal{V}$ of algebras has the property that the congruence lattices of its members consist of permuting equivalence relations if and only if $\mathcal{P} \leq \mathcal{V}$ where $\mathcal{P}$ is the variety with one ternary basic operation symbol p that is axiomatized by the **Maltsev identities**:

$$p(x, y, y) \approx x \qquad \text{and} \qquad p(y, y, x) \approx x. \tag{1.7}$$

In other words, $\mathcal{V}$ consists of congruence permutable algebras if and only if there is a ternary $\mathcal{V}$-term p, called a **Maltsev term**, such that the identities (1.7) hold in $\mathcal{V}$. Thus, one may refine Birkhoff's earlier explanation of Dedekind's modularity result to: groups are congruence modular because groups have a Maltsev term $p(x, y, z) = xy^{-1}z$, so groups have permuting congruences, and lattices of permuting equivalence relations are modular.

Let $\mathcal{U}$ be a finitely presented[2] variety. The condition on a variety $\mathcal{V}$ that $\mathcal{U} \leq \mathcal{V}$ may be expressed as "there is a finite set of $\mathcal{V}$-terms corresponding to the basic operation symbols of $\mathcal{U}$ such that the finite set of identities corresponding to the axioms of $\mathcal{U}$ hold in $\mathcal{V}$." A condition of this type is called a **strong Maltsev condition**, and given a finitely presented variety $\mathcal{U}$ the class $\{\mathcal{V} \mid \mathcal{U} \leq \mathcal{V}\}$ is the **strong Maltsev class** that is **defined** by this condition. Given a descending sequence $\cdots \leq \mathcal{U}_2 \leq \mathcal{U}_1 \leq \mathcal{U}_0$ of finitely presented varieties, the class $\{\mathcal{V} \mid \exists n(\mathcal{U}_n \leq \mathcal{V})\}$ is the **Maltsev class** that is defined by this sequence, and the **Maltsev condition** associated to this sequence is the assertion that for some n there is a finite set of $\mathcal{V}$-terms corresponding to the basic operation symbols of $\mathcal{U}_n$ such that the finite set of identities corresponding to the axioms of $\mathcal{U}_n$ hold in $\mathcal{V}$.

After Maltsev published the condition from (1.7) defining the class of varieties with permuting congruences, Alden Pixley found in [**73**] a strong Maltsev condition defining the class of varieties with distributive and permuting congruences, B. Jónsson found in [**38**] a Maltsev condition defining the class of varieties with distributive congruence lattices, and Alan Day found in [**8**] a Maltsev condition defining the class of varieties with modular congruence lattices. These results (and many others like them) culminated in the theorem, obtained independently by Pixley [**74**] and Rudolf Wille [**79**], that if ε is any lattice identity, then the class of varieties whose congruence lattices satisfy ε is the intersection of countably many Maltsev classes. This proves, in particular, that if two varieties satisfy the same Maltsev conditions, then they satisfy the same congruence identities. This result of Pixley and Wille includes an algorithm for generating Maltsev conditions associated with congruence identities, paving the way for a deeper study of congruence identities.

J. B. Nation discovered in [**70**] that there exist lattice identities that are strictly weaker than the modular law when considered as lattice identities, but equivalent to the modular law when considered as congruence identities. That is, there exists a variety $\mathcal{W}$ of lattices that strictly contains all modular lattices, yet any variety $\mathcal{V}$ of algebras with the property that the congruence lattice of every algebra from $\mathcal{V}$ is found in $\mathcal{W}$ is actually congruence modular. Nation's Theorem brought attention

[2]That is, $\mathcal{U}$ is a finitely axiomatized variety with finitely many basic operation symbols.

to the possibility that there might be "few" different **congruence varieties**, which are varieties of lattices of the form

$$\mathrm{CON}(\mathcal{V}) = \mathsf{HSP}\big(\{\mathbf{Con}(\mathbf{A}) \mid \mathbf{A} \in \mathcal{V}\}\big)\,. \tag{1.8}$$

For example, it suddenly became plausible that the only proper nondistributive congruence varieties might be those of the form $\mathrm{Con}(\mathcal{M})$ where $\mathcal{M}$ is a variety of modules. Nation's result led Ralph McKenzie to conjecture in [**66**] that if ε is any nontrivial lattice identity, then ε implies modularity when considered as a congruence identity. Studies of identities of small complexity yielded positive evidence for McKenzie's Conjecture in [**9, 10, 21, 39, 69**]. The most general of these results is the theorem of Ralph Freese and J. B. Nation that any lattice identity ε that can be written as an inclusion of the form

$$\bigwedge\bigvee\bigwedge \text{variables} \leq \bigvee\bigwedge\bigvee \text{variables}$$

implies modularity as a congruence identity. Similar results were proved at the same time about congruence identities implying congruence distributivity, and in [**17**] one finds the result that any congruence identity implying congruence modularity must also imply the stronger Arguesian identity as a congruence identity.

McKenzie's Conjecture was refuted by S. V. Polin in his famous paper [**75**]. Polin constructed a locally finite variety $\mathcal{P}$ that is not congruence modular, but satisfies a nontrivial congruence identity. To describe a congruence identity that holds in Polin's variety, let x, y and z be lattice variables and let $y_1 = y$, $z_1 = z$, $y_{n+1} = y + (x \cap z_n)$ and $z_{n+1} = z + (x \cap y_n)$. Let δ_n be the weakened distributive law:

$$x \cap (y + z) \approx (x \cap y_n) + (x \cap z_n)\,.$$

The identity δ_1 is the usual distributive law. Although δ_2 is strictly weaker than distributivity as a lattice identity, it implies distributivity as a congruence identity.[3] Polin's variety satisfies δ_3 as a congruence identity.[4] Since it does not satisfy δ_2, and since any lattice variety that satisfies δ_3 and not δ_2 is nonmodular, $\mathrm{Con}(\mathcal{P})$ is an example of a proper, nondistributive congruence variety that differs from $\mathrm{Con}(\mathcal{M})$ for any variety of modules $\mathcal{M}$.[5] Polin's variety was thoroughly investigated by Day and Freese in [**11**] with the result that we now have an efficient algorithm for determining if an identity implies modularity as a congruence identity. In an unpublished manuscript, Day showed that the Polin construction can be iterated[6] to produce varieties satisfying weaker and weaker nontrivial congruence identities.

Not all of the preceding results were originally proved via Maltsev conditions, but they could have been, and the analysis of Maltsev conditions is a powerful method for obtaining further results like these. Moreover, there are results on congruence identities that have been proved by a careful analysis of the Maltsev conditions defining a given class of varieties that seem unreachable by any other

[3] We know this because Ralph Freese showed us that δ_2 fails in $\mathrm{Con}(\mathbf{F}_{\mathcal{P}}(1))$. Using this observation and the main result of [**11**] it is easy to see that δ_2 implies distributivity as a congruence identity.

[4] Polin showed that $\mathcal{P}$ satisfies a different congruence identity. Alan Day was the one to emphasize the importance of the identities δ_n.

[5] We now know that there are continuum many different nonmodular congruence varieties (see [**36**]), and even some congruence varieties of groups that differ from any congruence variety of modules (see [**72**]).

[6] This and other modifications of Polin's construction were rediscovered and appear in [**54**] and [**71**].

method, such as Paolo Lipparini's results on congruence identities satisfied by congruence n-permutable varieties[7] in [**56, 58, 60, 61, 62**].

1.3. Commutator Theories

Birkhoff's HSP Theorem asserts that if $\mathcal{K}$ is a class of similar algebras, then any model of the identities true in $\mathcal{K}$ may be constructed as a homomorphic image of a subalgebra of a product of algebras in $\mathcal{K}$, that is as $\mathbf{B}/\theta$ where $\mathbf{B} \leq \prod \mathbf{A}_i$, $\mathbf{A}_i \in \mathcal{K}$. Replacing each $\mathbf{A}_i$ by the projection of $\mathbf{B}$ onto the i-th factor, we find that a typical model has the form $\mathbf{B}/\theta$ where $\mathbf{B}$ is a subdirect product of subalgebras of members of $\mathcal{K}$. It is therefore important to know how to construct congruences on subdirect products. The case of two factors is already difficult. If $\mathbf{B} \leq_{\mathrm{sd}} \mathbf{A}_1 \times \mathbf{A}_2$ and $\alpha_i \in \mathrm{Con}(\mathbf{A}_i)$, then the restriction of $\alpha_1 \times \alpha_2$ to $\mathbf{B}$ is a **product congruence** on $\mathbf{B}$. All other congruences are **skew**. For example, the congruence lattice of the

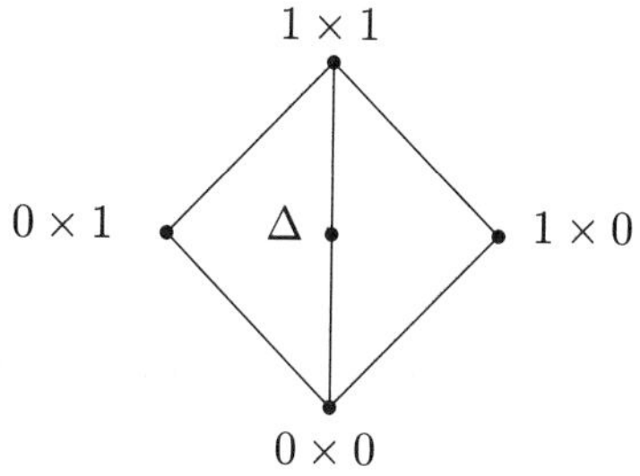

FIGURE 1.2. $\mathbf{Con}(\mathbb{Z}_2 \times \mathbb{Z}_2)$

group $\mathbf{B} = \mathbb{Z}_2 \times \mathbb{Z}_2$ is pictured in Figure 1.2. There is one skew congruence, Δ, which is the congruence that has the diagonal subgroup of $\mathbb{Z}_2 \times \mathbb{Z}_2$ as a class.

To understand skew congruences like this one, let us attempt to describe a typical "diagonal congruence" on a typical "diagonal subalgebra" $\mathbf{B}$ of $\mathbf{A}^2$ for some algebra $\mathbf{A}$ in a congruence permutable variety $\mathcal{V}$. Here a "diagonal subalgebra" is one containing the diagonal of $\mathbf{A}^2$, and a "diagonal skew congruence" is a congruence generated by a set of pairs of elements from the diagonal. More formally, if $\delta\colon \mathbf{A} \to \mathbf{A}^2\colon a \mapsto (a,a)$ is the canonical diagonal embedding, then $\mathbf{B} \leq \mathbf{A}^2$ is a diagonal subalgebra if δ factors through the inclusion of $\mathbf{B}$ into $\mathbf{A}^2$. A diagonal congruence on $\mathbf{B}$ is the extension to $\mathbf{B}$ of a congruence on $\delta(\mathbf{A})$.

Since we have assumed that $\mathbf{A}$ lies in a congruence permutable variety, it has a Maltsev term. Hence the only diagonal subalgebras $\mathbf{B}$ of $\mathbf{A}^2$ are those whose universe B is a congruence on $\mathbf{A}$, say $B = \beta$. Since the diagonal of $\mathbf{A}^2$ supports a subalgebra isomorphic to $\mathbf{A}$, the only diagonal congruences on $\mathbf{B}$ are the extensions to $\mathbf{B}$ of congruences $\delta(\alpha)$ for $\alpha \in \mathrm{Con}(\mathbf{A})$. Thus, the construction of a typical diagonal congruence on a typical diagonal subalgebra $\mathbf{B} \leq \mathbf{A}^2$ involves a pair of congruences $\alpha, \beta \in \mathrm{Con}(\mathbf{A})$. Specifically, we are considering $\mathbf{B}$ to be the subalgebra supported by β, and a diagonal congruence $\Delta = \Delta_{\alpha,\beta}$ on $\mathbf{B}$ that is generated by $\{\langle (u,u),(v,v)\rangle \mid (u,v) \in \alpha\}$. We next try to understand how Δ is related to the product congruences of $\mathbf{B}$.

[7]Congruences α and β are said to **n-permute** if the alternating composition $\alpha \circ_n \beta = \alpha \circ \beta \circ \alpha \cdots$, with $n-1$ occurrences of $\circ$, equals $\beta \circ_n \alpha$. A variety is **congruence n-permutable** if the congruences on all members n-permute.

Since we have assumed that **A** has a Maltsev term, and **B** is in the variety generated by **A**, **Con**(**B**) is modular. Dedekind's analysis of 3-variable consequences of the modular law shows that the 3-generated free modular lattice has only 28 elements, hence any 3-generated modular lattice can be easily drawn. In particu-

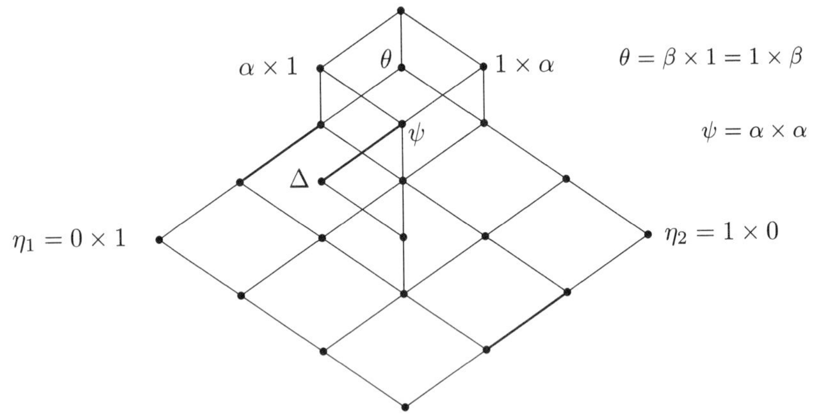

FIGURE 1.3. $\langle \eta_1, \eta_2, \Delta \mid \eta_1 \cap \eta_2 = 0 \rangle$

lar, the modular lattice with the presentation $\langle \eta_1, \eta_2, \Delta \mid \eta_1 \cap \eta_2 = 0 \rangle$ is the one in Figure 1.3. Hence the sublattice of **Con**(**B**) that is generated by the projection kernels $\eta_1 = 0 \times 1$ and $\eta_2 = 1 \times 0$ together with $\Delta = \Delta_{\alpha,\beta}$ is a homomorphic image of this lattice. The smallest product congruence containing Δ, the "product cover" of Δ, is

$$\big(\Delta + (0 \times 1)\big) \cap \big(\Delta + (1 \times 0)\big) = \alpha \times \alpha = \psi\,.$$

We take the interval $I[\Delta, \psi]$, from Δ to its product cover, to be a measure of the skewness of Δ. Meeting with η_2 and joining with η_1 projects this interval in two steps to the isomorphic interval

$$I\big[\eta_1 + (\Delta \cap \eta_2),\ \eta_1 + (\psi \cap \eta_2)\big] = I\big[\eta_1 + (\Delta \cap \eta_2),\ (\alpha \cap \beta) \times 1\big], \tag{1.9}$$

so the interval in (1.9) can also be taken to be a measure of the skewness of Δ. But the interval in (1.9) lies entirely in the interval above the first coordinate projection kernel η_1, which is naturally isomorphic to **Con**(**A**), so we can measure the skewness of Δ by considering the corresponding congruence interval of **A**. The

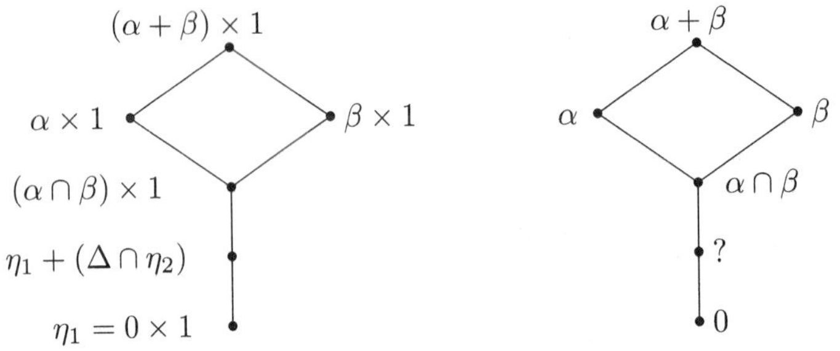

FIGURE 1.4. $I[0 \times 1, 1 \times 1]$ vs. **Con**(**A**)

left lattice in Figure 1.4 indicates the interval in **Con(B)** above η_1, while the right lattice shows the corresponding congruences on **A**. Note especially the question mark by the right lattice, which labels the congruence on **A** induced by Δ. This induced congruence is a function of α and β, and will be written $[\alpha, \beta]$ and called the **commutator** of α and β. For groups it is the ordinary group commutator.[8] The interval from the commutator $[\alpha, \beta]$ to the intersection $\alpha \cap \beta$ is the interval in **Con(A)** that measures the skewness of Δ.

In analogy with group theory, we call a congruence interval $I[\sigma, \tau]$ **abelian** if $[\tau, \tau] \leq \sigma$, and call the algebra **A** abelian if the entire lattice $\mathbf{Con(A)} = [0, 1]$ is an abelian congruence interval. If τ is defined to be equal to $\alpha \cap \beta$ in the last paragraph, then $\tau \leq \alpha$ and $\tau \leq \beta$ so $[\tau, \tau] \leq [\alpha, \beta]$. This shows that the interval $I\big[[\alpha, \beta], \alpha \cap \beta\big]$, which measures the skewness of Δ, is abelian. Conversely, every abelian interval $I[\sigma, \tau]$ is contained in an interval that measures the skewness of some diagonal congruence, namely $\Delta_{\tau,\tau}$. Hence diagonal congruences fail to be skew in $\mathcal{V}$ if and only if $\mathcal{V}$ omits abelian congruence intervals if and only if the commutator trivializes throughout $\mathcal{V}$ in the sense that

$$\forall \alpha, \beta\big([\alpha, \beta] = \alpha \cap \beta\big) \tag{1.10}$$

holds throughout $\mathcal{V}$. It can be shown that for varieties with a Maltsev term the commutator identity (1.10) is equivalent to congruence distributivity, and from congruence distributivity one can derive that there are no skew congruences of any type in finite subdirect products of algebras in $\mathcal{V}$. Thus, although it seems that we examined only a very special instance of a diagonal skew congruence, which then resulted in the definition of the commutator, this commutator controls all skew congruences in finite subdirect products in varieties with a Maltsev term.

In fact, the assumption that **A** has a Maltsev term was used only to deduce that every diagonal subalgebra of $\mathbf{A}^2$ has a congruence as its universe. This fact turns out to be inessential; the theory can be developed under the weaker hypothesis that **A** lies in a congruence modular variety. The theory of this commutator was developed by Jonathan D. H. Smith (when $\mathcal{V}$ has a Maltsev term) in [**76**], Joachim Hagemann and Christian Herrmann in [**28**], Heinz-Peter Gumm in [**27**], and Freese and McKenzie in [**19**]. The strength of the theory lies in the representation theorem for abelian algebras, proved by Herrmann in [**31**], which states that an abelian algebra in a congruence modular variety is **affine**. This theorem associates to an abelian algebra or congruence interval a module in the modern sense, i.e., a module over a ring. The fact that the commutator links every skew congruence with a module is the key to dealing with these congruences.

To emphasize the link with what we have said earlier, the commutator in congruence modular varieties encodes the existence of skew congruences. When the commutator trivializes, skew congruences are omitted, and this restricts the shape of the congruence lattices in some way. In the case of the modular commutator, this restriction on shapes is expressible by a congruence identity stronger than modularity, namely the distributive law.

When $\mathcal{V}$ is not congruence modular the types of skew congruences multiply. Different commutator theories have been invented which deal with this. The **TC-commutator** is a commutator based on the **term condition** (Section 2.5). It was

[8] This means that if N_α and N_β are the normal subgroups corresponding to the subscripted congruences, then $[N_\alpha, N_\beta] = N_{[\alpha,\beta]}$.

invented to generalize the modular commutator. As such it encodes the existence of diagonal skew congruences. A useful and completely general representation theorem for abelian algebras and congruences is not likely to exist, but Keith Kearnes and Ágnes Szendrei extended Herrmann's representation theorem for congruence modular varieties to any variety satisfying some nontrivial idempotent Maltsev condition in [**53**].

McKenzie next introduced the **strong term condition** in [**67, 34**], a concept similar in spirit to the ordinary term condition. He proved its usefulness in many ways, but he did not develop a commutator theory for it. Later McKenzie defined the concept of **rectangulation** (see [**55**]), and established its usefulness. His definition was slightly short of a term condition for rectangulation, and he did not develop a corresponding commutator theory.

In the early 1990's, we (= Kearnes and Kiss) worked on the problem of developing a commutator theory for McKenzie's strong term condition. We found that certain "cross-diagonal" skew congruences occur naturally in algebras. These are congruences on symmetric diagonal subalgebras $\mathbf{B} \leq \mathbf{A}^2$ generated by pairs of the form $\langle (u,v),(v,u)\rangle$. The behavior of this type of skew congruence is encoded in a commutator we call the **strong rectangular commutator**. It is so named because it has an associated term condition related to McKenzie's description for rectangulation. We developed the theory of this commutator in [**49**] and proved a representation for its abelian algebras and congruences. We found that McKenzie's strong term condition was exactly the conjunction of his original term condition (TC) and our term condition for strong rectangulation, therefore a commutator theory for the strong term condition follows from the theories of the TC-commutator and the strong rectangular commutator without further effort.

In this monograph we introduce a term condition for rectangulation, develop the theory of the rectangular commutator, and prove a representation theorem for its abelian algebras and congruences (Chapter 5).

There are actually many more commutators than the four described above, but the others that are known seem to be approximations to these four. Each of these commutators has an ideal model of an abelian algebra. For the TC commutator, the ideal model is an abelian group expanded by unary endomorphisms, i.e., a module over a ring. For the strong commutator it is a set expanded by unary endomorphisms, i.e., a unary algebra. For the rectangular commutator it is a semilattice expanded by unary endomorphisms, which is a natural type of semimodule. For the strong rectangular commutator it is a reduct to an antichain of a semilattice expanded by unary endomorphisms. The strangeness of this fourth notion of 'abelianness' is a reminder that these commutator theories do not begin by postulating the structure of an ideal abelian algebra, but rather by identifying a natural type of skew congruence.

1.4. The Results of this Monograph

The theorem which prompted us to write this monograph was our 1999 discovery that a variety satisfies a nontrivial congruence identity if and only if it satisfies an idempotent Maltsev condition that fails in the variety of semilattices (Theorem 7.15). David Hobby and Ralph McKenzie proved in Chapter 9 of [**34**] that a certain Maltsev condition, that we will call in this introduction "HM", defines the class of all varieties satisfying an idempotent Maltsev condition that fails in the variety of semilattices. HM is a disjunction $\bigvee \mathrm{HM}_n$ of strong Maltsev conditions. The proof of the 1999 theorem is based on a careful analysis of various Maltsev conditions equivalent to HM. A novel feature of the proof is that it deals only with the local effects of HM on nonsolvable intervals in congruence lattices of algebras in a variety $\mathcal{V}$ satisfying some HM_n. Namely, we prove that every nonmodular interval in a congruence lattice contains what we call a "solvability obstruction". We use HM_n to introduce a rank function on solvability obstructions that appear in congruence lattices of algebras in $\mathcal{V}$. Then we prove that for every solvability

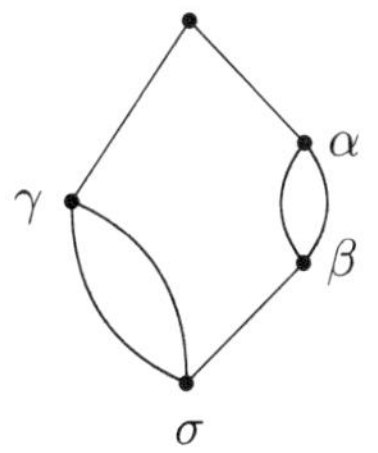

FIGURE 1.5

obstruction that appears in the interval $I[\beta, \alpha]$ of some copy of $\mathbf{N}_5$ in some congruence lattice there is associated a solvability obstruction of strictly smaller rank in the interval $I[\sigma, \gamma]$ (cf. Figure 1.5). By iterating this observation, one obtains that

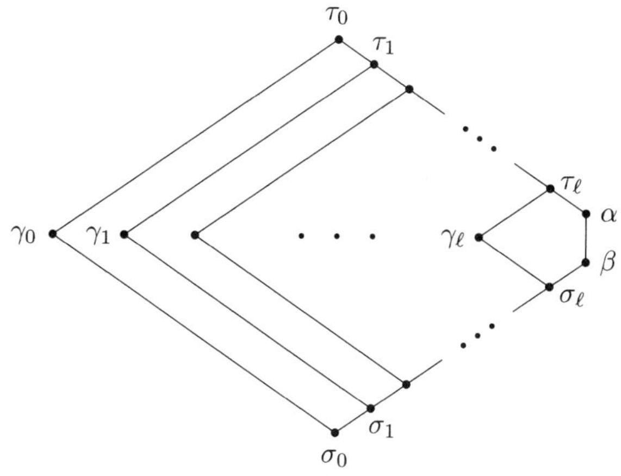

FIGURE 1.6. The lattice $\mathbf{N}_{\ell+5}$

deeply nested copies of $\mathbf{N}_5$, as depicted in Figure 1.6, can appear as a sublattice of a congruence lattice of an algebra in $\mathcal{V}$ only if solvability obstructions of large rank exist. We complete the proof by proving that the complexity of HM_n induces a uniform (finite) bound on the rank of solvability obstructions throughout $\mathcal{V}$, hence for some ℓ it is not possible to embed $\mathbf{N}_{\ell+5}$ into the congruence lattice of any member of $\mathcal{V}$. This fact is converted into a nontrivial congruence identity satisfied by $\mathcal{V}$.[9]

[9]It is interesting that Dedekind's investigation into the arithmetic of congruences of $\langle \mathbb{C}; +, -, 0\rangle$ led to the law postulating the omission of sublattices isomorphic to $\mathbf{N}_5$, while the natural extension of that investigation to arbitrary algebras leads to "generalized modular laws" postulating the omission of sublattices isomorphic to $\mathbf{N}_{\ell+5}$ for some ℓ.

This theorem gives a simple answer to the question "Which varieties satisfy nontrivial congruence identities?" But more importantly, the effort to prove the theorem answers the questions "Which varieties have trivial rectangular commutator?" and "What kinds of restrictions on congruence lattice shapes are linked to having trivial rectangular commutator?", since the combination of Theorems 5.25 and 7.15 prove that a variety $\mathcal{V}$ satisfies a nontrivial congruence identity if and only if the rectangular commutator is trivial throughout $\mathcal{V}$.

This monograph contains many characterizations of the class of varieties whose rectangular commutator is trivial: by the shapes of congruence lattices in these varieties and by the Maltsev conditions they satisfy (see, for example, Sections 5.3, 7.2 and 8.2). But it also contains characterizations of other classes of varieties with parallel descriptions. Namely, we consider the classes of varieties (i) whose strong commutator is trivial (Sections 3.1, 4.2 and 4.3), (ii) whose strong rectangular commutator is trivial (Sections 3.2, 4.2 and 4.3), (iii) whose TC commutator is trivial (Section 8.1), and (iv) whose rectangular and TC commutators are both trivial (Section 8.3). These characterizations of classes of varieties are also in terms of the shapes of congruence lattices in these varieties and the Maltsev conditions they satisfy. One byproduct of this investigation is our first real understanding of congruence join semidistributivity: a variety $\mathcal{V}$ is congruence join semidistributive if and only if it is congruence meet semidistributive and satisfies a nontrivial congruence identity (equivalently, if all commutators trivialize throughout $\mathcal{V}$).

We end the monograph with a chapter on residually small varieties satisfying congruence identities (Chapter 9). The main result of this chapter is that a residually small variety satisfies a congruence identity if and only if it is congruence modular, hence McKenzie's Conjecture is true for residually small varieties. We apply the main result of the chapter to show that there is no "almost congruence distributive" variety that satisfies a nontrivial idempotent Maltsev condition.

A number of problems are posed in this monograph. Those posed in the text are restated after Chapter 9 (see page 155), together with some new problems.

We attach an appendix which collects the many characterizations from the monograph of the classes of varieties that have Taylor terms or Hobby-McKenzie terms.

1.5. Thanks!

We are happy to express our gratitude to those who supplied comments, corrections and encouragement during the preparation of this monograph, including: Joel Berman, Ralph Freese, Paweł Idziak, Bill Lampe, Paolo Lipparini, Ralph McKenzie, George McNulty, J. B. Nation, Peter Ouwehand, James Raftery, Ágnes Szendrei, Walter Taylor, Jirí Tuma, Matthew Valeriote, Fred Wehrung, and Ross Willard.

CHAPTER 2

Preliminary Notions

Here we fix notation and introduce definitions and results with the purpose of providing a bridge from standard material to the material in this monograph.

In this chapter and the sequel we use the following conventions concerning notation. The set of natural numbers is ω. The first natural number is 0. Sequences or tuples of elements from a set X are written in boldface, as in $\mathbf{x} \in X$ (or $\mathbf{x} \in X^n$ if the length n is to be specified). The i-th coordinate of the tuple $\mathbf{x}$ is written in italic as x_i. If two sequences $\mathbf{x}, \mathbf{y} \in X$ have the same length and R is a binary relation on X, then we may write $\mathbf{x} \equiv \mathbf{y} \pmod{R}$ or $\mathbf{x} \equiv_R \mathbf{y}$ or $\mathbf{x}\ R\ \mathbf{y}$ to mean that $(x_i, y_i) \in R$ for all i. If R is a binary relation on X, then the set $X^n[R]$ consists of the tuples $\mathbf{x} \in X^n$ such that $x_i \equiv_R x_j$ for all $1 \leq i \leq j \leq n$. The projection of a subset $S \subseteq \prod_{i \in I} X_i$ onto a subset of coordinates $J \subseteq I$ is denoted $\pi_J \colon S \to \prod_{j \in J} X_j$, and its kernel is denoted η_J. If θ is an equivalence relation on $\prod_{j \in J} X_j$, then θ_J denotes $\pi_J^{-1}(\theta)$ (so $0_J = \eta_J$). When $J = \{j\}$, then we write π_j, η_j and θ_j instead of $\pi_{\{j\}}$, $\eta_{\{j\}}$ and $\theta_{\{j\}}$. Expressions $A := B$ or $B =: A$ mean "A is defined by B".

2.1. Algebras, Varieties, and Clones

An algebra is a model of a 1-sorted first-order algebraic language. To fix conventions, an **algebraic signature** is a pair $\sigma := (F, \alpha)$ where F is a set (of operation symbols), and $\alpha \colon F \to \omega$ is a function (assigning arity). An **algebra** of signature σ is a pair $\mathbf{A} := \langle A; F \rangle$ where A is a nonempty set, called the **universe** of $\mathbf{A}$, and for each $f \in F$ with $\alpha(f) = k$ there is an assigned k-ary operation $f^{\mathbf{A}} \colon A^k \to A$. The operations $f^{\mathbf{A}}$ are called the **basic operations** of $\mathbf{A}$.

If Y is a set that is disjoint from F, then $T(Y)$ is the smallest set of words in the alphabet $F \cup Y$ satisfying

(i) $Y \subseteq T(Y)$, and
(ii) if $f \in F$, $\alpha(f) = k$ and $g_1, \ldots, g_k \in T(Y)$, then the word $fg_1 \cdots g_k \in T(Y)$.

When $Y = X := \{x_i \mid 1 \leq i < \omega\}$, then $T(X)$ is the set of **terms of signature** σ, and its members are called **terms**. Similarly, if $Y = X_n := \{x_1, \ldots, x_n\}$, the members of $T(X_n)$ are **n-ary terms**.

The set $T(Y)$ has a canonical structure of an algebra of signature σ. Namely, if $f \in F$ has arity k, then in the term algebra $\mathbf{T} = \mathbf{T}(Y)$ the operation $f^{\mathbf{T}}$ is defined so that if $g_1, \ldots, g_k \in T$, then $f^{\mathbf{T}}(g_1, \ldots, g_k) = fg_1 \cdots g_k \in T$. If $\mathbf{A}$ is an algebra of signature σ and $Y = X$ or X_n, then an **assignment in A** of the variables Y is a function $v \colon Y \to A$. It follows from the unique readability of terms that any assignment v extends uniquely to an algebra homomorphism $\widehat{v} \colon \mathbf{T} \to \mathbf{A}$; that is, $\mathbf{T}$ is free over Y in the class of all algebras of signature σ. If $t \in T$, then both $\widehat{v}(t)$ and $t(v)$ denote the image of t under $\widehat{v}$.

If $\mathbf{A}$ is an algebra of signature σ and $t \in T(X_n)$, then t determines an n-ary **term operation** on A, the function $t^{\mathbf{A}}\colon A^n \to A$ is defined by $(a_1, \ldots, a_n) \mapsto t(v)$ where $v\colon X_n \to A$ is the assignment $x_i \mapsto a_i$. These remarks apply in the case where $\mathbf{A} = \mathbf{T} = \mathbf{T}(X_n)$, and show that, if $t \in T(X_k)$ and $g_1, \ldots, g_k \in T(X_n)$, then there is a term $t(g_1, \ldots, g_k) := t^{\mathbf{T}}(g_1, \ldots, g_k) \in T(X_n)$. It can be proved by induction that, in this notation, the term $t(x_1, \ldots, x_k)$ is simply t itself.

The **language** L associated with the signature σ is the set of first-order formulas in this signature. Parentheses may be added to formulas to improve readability. An **identity** in the language L is an atomic L-formula, which is a formula of the form $p \approx q$ where $p, q \in T(X_n)$ for some n. An algebra $\mathbf{A}$ **satisfies** $p \approx q$, written $\mathbf{A} \models p \approx q$, if $p(v) = q(v)$ for every assignment in $\mathbf{A}$ (equivalently if $p^{\mathbf{A}} = q^{\mathbf{A}}$). A class of algebras in the language L is a **variety** if it is definable as the class of all L-algebras satisfying some set of identities. The smallest variety containing an algebra $\mathbf{A}$ is denoted $\mathcal{V}(\mathbf{A})$.

A **quasi-identity** is an open first-order formula of the form

$$\bigwedge_{i=1}^{n} (p_i \approx q_i) \to (p \approx q)$$

where $p \approx q$ and each $p_i \approx q_i$ is an identity. A class of L-algebras is a **quasi-variety** if it is definable by some set of quasi-identities.

A **clone** is a multisorted structure $\mathcal{C} = \langle C_0, C_1, \ldots; F\rangle$, with sorts indexed by ω, where

(i) Each of the sorts C_i, $i \in \omega$, is a set.
(ii) $F = \{\pi_i^n \mid n \in \omega, 1 \le i \le n\} \cup \{\mathrm{comp}_n^m \mid m, n \in \omega\}$ is a set of operations between the C_i.
(iii) Each π_i^n is a 0-ary operation (a constant) in C_n.
(iv) $\mathrm{comp}_n^m\colon C_m \times (C_n)^m \to C_n$ is an $(m+1)$-ary operation.
(v) $\mathrm{comp}_n^n(f, \pi_1^n, \ldots, \pi_n^n) = f$ and $\mathrm{comp}_n^m(\pi_i^m, f_1, \ldots, f_m) = f_i$.
(vi) $\mathrm{comp}_m^p\big(f, \mathrm{comp}_m^n(g_1, h_1, \ldots, h_n), \ldots, \mathrm{comp}_m^n(g_p, h_1, \ldots, h_n)\big)$
$= \mathrm{comp}_m^n\big(\mathrm{comp}_n^p(f, g_1, \ldots, g_p), h_1, \ldots, h_n\big)$.

For an example of a clone, take C_n to be the set of n-ary L-terms, take π_i^n to be x_i, and take comp_n^m to be the composition of an m-ary term with m n-ary terms. This is the **clone of L-terms**. The **clone of A** when $\mathbf{A}$ is an algebra is defined similarly using the term operations of $\mathbf{A}$ instead of the L-terms.

It is evident that clones are multisorted algebras defined by identities, hence the class of clones is a multisorted variety. This implies that the usual algebraic notions apply to clones. In particular, a sequence

$$\mathbf{h} = (h_0, h_1, \ldots)\colon \mathcal{C} \to \mathcal{D}$$

is a **homomorphism** between clones if each $h_i\colon C_i \to D_i$ is a function, and the sequence $\mathbf{h}$ preserves the clone operations of composition and projection. The notions of kernel and quotient are defined in the obvious way.

Let $\mathcal{V}$ be a variety of 1-sorted algebras in the language L. The **clone of $\mathcal{V}$**, denoted $\mathrm{Clo}(\mathcal{V})$, is that quotient of the clone of L-terms that is obtained by identifying terms $p, q \in C_n$ if $p \approx q$ is satisfied by all algebras in $\mathcal{V}$. The correspondence $\mathcal{V} \mapsto \mathrm{Clo}(\mathcal{V})$ is essentially bijective. Namely, each variety $\mathcal{V}$ is assigned the clone $\mathrm{Clo}(\mathcal{V})$, and each clone $\mathcal{C}$ is assigned a variety $\mathrm{Var}(\mathcal{C})$ which we define now. The operation symbols of $\mathrm{Var}(\mathcal{C})$ are $\bigcup_{i\in\omega} C_i$ where the arity of $f \in C_n$ is defined to be n. The identities defining $\mathrm{Var}(\mathcal{C})$ are of two types:

(i) projection identities: $\pi_i^n(x_1,\dots,x_n) \approx x_i$ for each $1 \le i \le n$, and
(ii) composition identities:

$$f(x_1,\dots,x_n) \approx g\big(h_1(x_1,\dots,x_n),\dots,h_m(x_1,\dots,x_n)\big)$$

whenever the equality $f = \mathrm{comp}_n^m(g,h_1,\dots,h_m)$ holds in $\mathcal{C}$.

It can be shown that $\mathcal{C} = \mathrm{Clo}\big(\mathrm{Var}(\mathcal{C})\big)$ holds for any clone $\mathcal{C}$, while $\mathcal{V}$ is definitionally equivalent to the variety $\mathrm{Var}\big(\mathrm{Clo}(\mathcal{V})\big)$ for any variety $\mathcal{V}$.

We write $\mathcal{U} \le \mathcal{V}$ if there is a homomorphism $\mathbf{h}\colon \mathrm{Clo}(\mathcal{U}) \to \mathrm{Clo}(\mathcal{V})$. This notation is used to express the fact that the algebras in $\mathcal{V}$ have an underlying $\mathcal{U}$-structure.

2.2. Lattice Theory

A **partial lattice** is a structure $\mathbf{P} = \langle P; \vee, \wedge, \le\rangle$ where $\langle P; \le\rangle$ is a partially ordered set and $\vee$ and $\wedge$ are partial binary operations on P (called **join** and **meet** respectively) such that, if $a \vee b$ is defined, then it is the least upper bound of a and b in $\langle P; \le\rangle$, and if $a \wedge b$ is defined, then it is the greatest lower bound of a and b in $\langle P; \le\rangle$. A **lattice** is a total algebra $\mathbf{L} = \langle L; \vee, \wedge\rangle$ on which there is a partial order $\le$ making $\langle L; \vee, \wedge, \le\rangle$ a partial lattice.

If $x \le y$ in some lattice $\mathbf{L}$, then the **interval** between them is $I[x,y] := \{z \mid x \le z \le y\}$. A **lattice ideal** of $\mathbf{L}$ is a subset $I \subseteq L$ closed under join ($x, y \in I \implies x \vee y \in I$) and closed downward ($x \in I \;\&\; z \le x \implies z \in I$). The dual concept is a **lattice filter**. The **principal ideal** determined by $x \in L$ is $(x] := \{z \mid z \le x\}$. The collection $\mathcal{I}(\mathbf{L})$ of all ideals of $\mathbf{L}$, ordered by inclusion, is the **ideal lattice** of $\mathbf{L}$. If $I, J \in \mathcal{I}(\mathbf{L})$, then $I \wedge J = I \cap J$ and

$$I \vee J = \big\{z \mid \exists x \in I, y \in J (z \le x \vee y)\big\}\,.$$

Lattice terms may be called **words**. If p and q are n-ary lattice words, then the inclusion $p \le q$ is **satisfied** by a lattice $\mathbf{L}$ if and only if $p^{\mathbf{L}} \le q^{\mathbf{L}}$ in the pointwise order. Therefore $\mathbf{L}$ satisfies $p \approx q$ if and only if it satisfies both $p \le q$ and $q \le p$. Note also that $p \le q$ is satisfied by $\mathbf{L}$ if and only if the identity $p \approx p \wedge q$ is satisfied by $\mathbf{L}$. It follows that a class of lattices is definable by identities if and only if it is definable by inclusions. The variety of all lattices is denoted $\mathcal{L}$.

If Q is the lattice quasi-identity $\bigwedge(p_i \approx q_i) \to (p \approx q)$, then the Q**-configuration** is a pair $\big(\mathbf{P}(Q), p \approx q\big)$ where $\mathbf{P}(Q)$ is the natural partial lattice of subterms of terms of Q. That is, $\mathbf{P}(Q)$ is the partial lattice presented by $\langle G \mid R\rangle$ where G is the set of subterms of terms appearing in Q and R consists of relations of the following types: if s, t and $s \vee t$ are subterms, then R contains a relation expressing that $s \vee t$ is equal to the join of s and t, R contains similar relations for meet, and for each premise $p_i \approx q_i$ of Q the set R contains the relation $p_i = q_i$. It is not assumed that R contains $p = q$ where $p \approx q$ is the conclusion of Q. This definition exists to make the following statement true: an assignment of the variables of Q in a lattice $\mathbf{L}$ which satisfies the premises of Q determines and is determined by a homomorphism of partial lattices $\varphi\colon \mathbf{P}(Q) \to \mathbf{L}$, and the assignment will satisfy Q precisely when $\varphi(p) = \varphi(q)$.

DEFINITION 2.1. The **meet semidistributive law** is the quasi-identity

$$\big((p \wedge q) \approx s\big) \;\&\; \big((p \wedge r) \approx s\big) \to \big((p \wedge (q \vee r)) \approx s\big)\,, \tag{2.1}$$

and the **join semidistributive law** is the dual quasi-identity.

Quasi-identity (2.1) is equivalent to

$$\big((p \wedge q) \approx s\big) \mathbin{\&} \big((p \wedge r) \approx s\big) \mathbin{\&} \big((q \vee r) \geq p\big) \to (p \approx s)\,. \tag{2.2}$$

The reason this is so is that any assignment of variables which fails to satisfy (2.2) will also fail to satisfy (2.1), while conversely if $p \mapsto a$, $q \mapsto b$, $r \mapsto c$, $s \mapsto d$ is an assignment of variables in some lattice that fails to satisfy (2.1), then $p \mapsto a \wedge (b \vee c)$, $q \mapsto b$, $r \mapsto c$, $s \mapsto d$ is an assignment that fails to satisfy (2.2).

The $\mathrm{SD}_\wedge$**-configuration** is the Q-configuration where Q is quasi-identity (2.2). More explicitly, let $\mathbf{P}(\mathrm{SD}_\wedge)$ be the partial lattice generated by $\{p, q, r, m, j\}$ where $m = p \wedge q = p \wedge r$ and $j = q \vee r \geq p$. Then the $\mathrm{SD}_\wedge$-configuration is $\big(\mathbf{P}(\mathrm{SD}_\wedge), p \approx m\big)$. An $\mathrm{SD}_\wedge$**-failure** in a lattice $\mathbf{L}$ is an interval of the form $I = I[\varphi(m), \varphi(p)]$ where $\varphi\colon \mathbf{P}(\mathrm{SD}_\wedge) \to \mathbf{L}$ is a homomorphism of partial lattices. This $\mathrm{SD}_\wedge$-failure is **trivial** if I has one element and is **nontrivial** otherwise. **The** $\mathrm{SD}_\vee$**-configuration** and $\mathrm{SD}_\vee$**-failures** are defined dually. Thus a lattice is meet semidistributive if and only if it has no nontrivial $\mathrm{SD}_\wedge$-failures.

For finite lattices, or for varieties of lattices, the meet and join semidistributive laws are characterized in the following theorem.

THEOREM 2.2.

(1) (Cf. [**7**]) *A finite lattice is meet semidistributive if and only if it has no sublattice isomorphic to* $\mathbf{M}_3$*,* $\mathbf{D}_1$*,* $\mathbf{E}_1$*,* $\mathbf{E}_2$ *and* $\mathbf{G}$*. (These are five of the following six lattices.)*

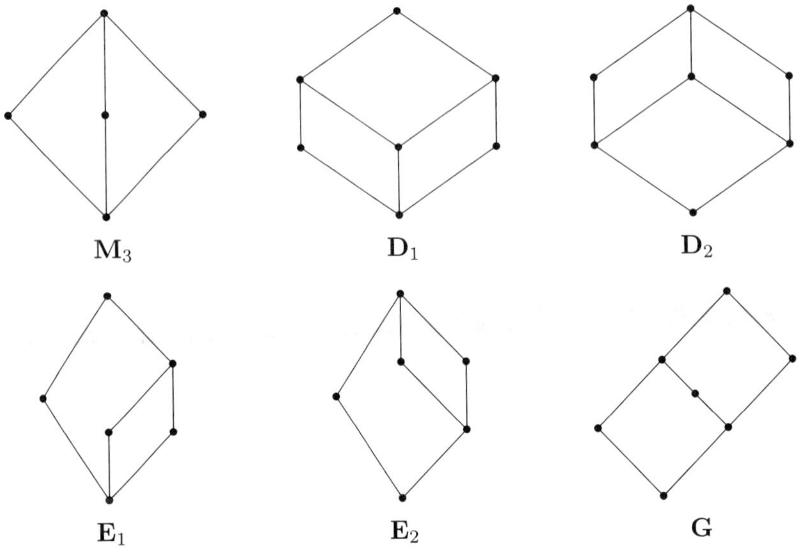

FIGURE 2.1

(2) (Cf. [**41**]) *A variety of lattices consists of meet semidistributive lattices if and only if it does not contain* $\mathbf{M}_3$*,* $\mathbf{D}_1$*,* $\mathbf{E}_1$*,* $\mathbf{E}_2$ *or* $\mathbf{G}$*.*

The dual results hold for join semidistributivity in place of meet semidistributivity.

DEFINITION 2.3. **Whitman's condition**, denoted by (**W**), is the formula

$$(x \wedge y \leq u \vee v) \rightarrow$$
$$\big((x \leq u \vee v) \text{ or } (y \leq u \vee v) \text{ or } (x \wedge y \leq u) \text{ or } (x \wedge y \leq v)\big).$$

This condition may be written a little more compactly as:

$$x \wedge y \leq u \vee v \quad \text{implies} \quad \{x, y, u, v\} \cap I[x \wedge y, u \vee z] \neq \emptyset.$$

All lattices depicted in Theorem 2.2 satisfy (W).

DEFINITION 2.4. A quasi-identity $\bigwedge_{i=1}^{n}(p_i \approx q_i) \rightarrow (p \approx q)$ in k variables **satisfies (W)** if the lattice with the presentation $\langle G \mid R\rangle$ satisfies (W), where $G = \{g_1, \ldots, g_k\}$, $\mathbf{g} = (g_1, \ldots, g_k)$ and $R = \{p_1(\mathbf{g}) = q_1(\mathbf{g}), \ldots, p_n(\mathbf{g}) = q_n(\mathbf{g})\}$.

Note that whether or not a quasi-identity satisfies (W) depends only on its premises. It is proved in [**6**] that a quasi-identity Q satisfies (W) if and only if the associated partial lattice $\mathbf{P}(Q)$ satisfies (W). Since $\mathbf{P}(Q)$ is finite, this yields an algorithm for testing a quasi-identity for (W). It is noted in [**6**] that the meet semidistributive law satisfies (W), and that more generally any quasi-identity whose premises are free of joins (as in form (2.1) of the meet semidistributive law) satisfies (W). These remarks also apply to the join semidistributive law.

DEFINITION 2.5. A surjective lattice homomorphism $h\colon \mathbf{K} \rightarrow \mathbf{L}$ is **upper bounded** if each kernel class $h^{-1}(a)$, $a \in L$, has a largest element, is **lower bounded** if each kernel class has a least element, and is **bounded** if it is both lower and upper bounded. A lattice $\mathbf{L}$ is **(upper, lower) bounded** if there is a surjective (upper, lower) bounded homomorphism $h\colon \mathbf{F}_{\mathcal{L}}(x_1, \ldots, x_n) \rightarrow \mathbf{L}$ from a finitely generated free lattice onto $\mathbf{L}$.

Any lower bounded lattice is join semidistributive (Theorem 2.20 of [**16**]). Chapter 2 of [**16**] describes an effective algorithm for testing if a finite lattice is lower or upper bounded. We shall not have cause to use this algorithm, but for later reference we point out that $\mathbf{D}_1$ is the only lower bounded lattice depicted in Theorem 2.2. The others are excluded because they are not join semidistributive, while the lower boundedness of $\mathbf{D}_1$ is proved one way in Example 2.74 of [**16**] and differently in Table 3 of [**65**].

DEFINITION 2.6. An algebra $\mathbf{P}$ is **projective** relative to a variety $\mathcal{V}$ if whenever $\sigma\colon \mathbf{A} \rightarrow \mathbf{B}$ is a surjective homomorphism between algebras of $\mathcal{V}$ and $\varphi\colon \mathbf{P} \rightarrow \mathbf{B}$ is a homomorphism, then there is a homomorphism $\overline{\varphi}\colon \mathbf{P} \rightarrow \mathbf{A}$ such that $\sigma \circ \overline{\varphi} = \varphi$.

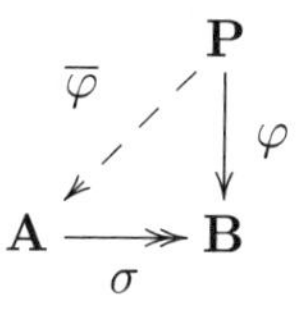

When $\mathcal{V}$ is a variety of lattices, we may call a partial lattice $\mathbf{P}$ **projective** relative to $\mathcal{V}$ if it satisfies the homomorphism lifting property of this definition.

THEOREM 2.7. *If $\mathcal{V}$ is a variety of algebras and $\mathbf{P}$ is subdirectly irreducible and projective relative to $\mathcal{V}$, then there is an identity ε such that for all $\mathbf{A} \in \mathcal{V}$ it is the case that $\mathbf{A} \models \varepsilon$ if and only if $\mathbf{A}$ has no subalgebra isomorphic to $\mathbf{P}$.*

We call the identity ε the **conjugate identity** for $\mathbf{P}$.

PROOF. Let $\mathcal{U}$ consist of all members of $\mathcal{V}$ that have no subalgebra isomorphic to $\mathbf{P}$. Then $\mathcal{U}$ is closed under the formation of homomorphic images because $\mathbf{P}$ is projective relative to $\mathcal{V}$, $\mathcal{U}$ is closed under subalgebras by definition, and $\mathcal{U}$ is closed under products because $\mathbf{P}$ is subdirectly irreducible. By Birkhoff's HSP Theorem, $\mathcal{U}$ is a subvariety of $\mathcal{V}$. Since $\mathbf{P} \notin \mathcal{U}$, there is an identity ε that is satisfied in $\mathcal{U}$ but not by $\mathbf{P}$. Necessarily ε fails in every member of $\mathcal{V} - \mathcal{U}$, so this identity has the property stated in the theorem. □

We shall have occasion to refer to the following closely-related result.

THEOREM 2.8. *If Q is a quasi-identity and $\mathbf{P}(Q)$ is projective relative to $\mathcal{V}$, then the class of lattices in $\mathcal{V}$ that satisfiy Q is a subvariety of $\mathcal{V}$.*

PROOF. As above, let $\mathcal{U}$ consist of all members of $\mathcal{V}$ that satisfy Q. Then $\mathcal{U}$ is closed under the formation of homomorphic images because $\mathbf{P}(Q)$ is projective relative to $\mathcal{V}$, $\mathcal{U}$ is closed under subalgebras and products because Q is a quasi-identity. □

2.3. Meet Continuous Lattice Theory

A lattice is **meet continuous** if it is complete and the binary meet operation distributes over arbitrary up-directed suprema. That is, for every up-directed set $\mathcal{D}$,

$$x \wedge \bigvee_{y \in \mathcal{D}} y = \bigvee_{y \in \mathcal{D}} x \wedge y \,. \tag{2.3}$$

Every algebraic lattice is meet continuous (Lemma VIII.5.2 of [**1**]).

The language of meet continuous lattices has operation symbols $\{\bigvee, \wedge\}$ where $\bigvee$ is a class of κ-ary join operations for all $\kappa \geq 0$ and $\wedge$ is binary meet. The 0-ary join is a constant that interprets as the least element and is denoted 0. The fact that $\wedge$ is a meet operation is expressible by identities in the same way that it is expressed in lattice theory. As with binary join, it is easy to express with identities the fact that κ-ary $\bigvee$ is κ-ary join with respect to the $\wedge$-order. To show that meet continuity is expressible by identities one must show that the distributive law in (2.3) can be rewritten with quantification over arbitrary sets rather than up-directed sets. This is done by replacing an arbitrary join with the up-directed supremum of its finite sub-joins:

$$x \wedge \bigvee_{y \in \mathcal{F}} y \approx \bigvee_{\substack{\mathcal{F}_0 \subseteq \mathcal{F} \\ \mathcal{F}_0 \text{ finite}}} \Big(x \wedge \bigvee_{y \in \mathcal{F}_0} y \Big).$$

The category $\mathcal{L}_{MC}$ of meet continuous lattices with join complete homomorphisms has a forgetful functor $F\colon \mathcal{L}_{MC} \to \mathcal{L}$ to lattices that forgets all join operations except the binary join. Theorem 2.9, which is essentially due to Ralph Freese, asserts that this functor has a left adjoint given by the ideal lattice functor. Here if $h\colon \mathbf{K} \to \mathbf{L}$ is a homomorphism between lattices, then $\mathcal{I}(h)\colon \mathcal{I}(\mathbf{K}) \to \mathcal{I}(\mathbf{L})$ is defined by

$$\mathcal{I}(h)(I) = \big\{ z \in L \mid \exists x \in I \big(z \leq h(x) \big) \big\} \,.$$

THEOREM 2.9. *Restriction to $\mathbf{L}$ is a natural bijection*

$$|\colon \mathcal{L}_{MC}\big(\mathcal{I}(\mathbf{L}), \mathbf{K}\big) \to \mathcal{L}\big(\mathbf{L}, F(\mathbf{K})\big).$$

(We are identifying $\mathbf{L}$ with the sublattice of $\mathcal{I}(\mathbf{L})$ consisting of principal ideals.)

PROOF. If $h\colon \mathcal{I}(\mathbf{L}) \to \mathbf{K}$ is a meet continuous lattice homomorphism, then the restriction $h|_{\mathbf{L}}\colon \mathbf{L} \to F(\mathbf{K})$ is evidently a lattice homomorphism (since lattices are reducts of meet continuous lattices). If $g\colon \mathbf{L} \to F(\mathbf{K})$ is a lattice homomorphism, then it is proved in Lemma 5.1 of [**15**] that the function

$$\widehat{g}\colon \mathcal{I}(\mathbf{L}) \to \mathbf{K}\colon I \mapsto \bigvee\{g(x) \mid x \in I\}$$

is a meet continuous lattice homomorphism whose restriction to $\mathbf{L}$ is g. It is the unique extension of g to $\mathcal{I}(\mathbf{L})$, since $\mathcal{I}(\mathbf{L})$ is generated under $\bigvee$ by L. This proves that restriction to $\mathbf{L}$ is a bijection between hom-sets. The naturality is left as an exercise. (See Chapter IV of [**64**] for the method.) □

The adjunction from $\mathcal{L}$ to $\mathcal{L}_{MC}$ may be composed with the adjunction from $\mathcal{SET}$ to $\mathcal{L}$ (given by the free lattice and forgetful functors) to produce a left adjoint $\mathcal{SET} \xrightarrow{free} \mathcal{L} \xrightarrow{\mathcal{I}} \mathcal{L}_{MC}$ to the composite forgetful functor $\mathcal{L}_{MC} \to \mathcal{L} \to \mathcal{SET}$. This adjoint establishes the existence and structure of free meet continuous lattices.

COROLLARY 2.10. *If $\mathbf{F} = \mathbf{F}_{\mathcal{L}}(X)$ is the free lattice generated by X, then $\mathcal{I}(\mathbf{F})$ is a free meet continuous lattice over the set $\{(x] \mid x \in X\}$.*

Since $\mathcal{L}_{MC}$ is definable by identities and has free algebras of all ranks, Birkhoff's HSP theorem is valid for $\mathcal{L}_{MC}$: a subclass of $\mathcal{L}_{MC}$ is definable by identities if and only if it is closed under the formation of homomorphic images, subalgebras and products. Hence the proof of Theorem 2.7 is valid for varieties of meet continuous lattices. Using known methods it can be shown that the projective subdirectly irreducible members of $\mathcal{L}_{MC}$ are the finite, lower bounded, subdirectly irreducible lattices satisfying (W). This includes all lattices that are projective and subdirectly irreducible in $\mathcal{L}$, along with some others (such as $\mathbf{D}_1$).

2.4. Maltsev Conditions

A **strong Maltsev condition** is a primitive positive sentence in the language of clones. This means that it is a first-order sentence of the form "$\exists \bigwedge$(atomic)" about clones. A **Maltsev condition** is a countably infinite disjunction $\bigvee_{i\in\omega} \sigma_i$ where the σ_i are strong Maltsev conditions that get weaker as i increases (i.e., $\sigma_i \vdash \sigma_{i+1}$ for all i). A variety **satisfies** a (strong) Maltsev condition if its clone does.

The concept of a Maltsev condition can be reformulated in terms of clone homomorphisms. To each strong Maltsev condition $\exists F \bigwedge \Sigma$ corresponds a finite presentation $\langle F \mid \Sigma\rangle$ of a clone. If $\mathcal{U}$ is a variety whose clone has this presentation, then $\mathcal{V}$ satisfies the strong Maltsev condition $\exists F \bigwedge \Sigma$ if and only if $\mathcal{U} \leq \mathcal{V}$. Similarly, to each Maltsev condition $\bigvee_{i\in\omega} \sigma_i$ corresponds a sequence $\cdots \leq \mathcal{U}_2 \leq \mathcal{U}_1 \leq \mathcal{U}_0$ of varieties with finitely presented clones. A variety $\mathcal{V}$ satisfies $\bigvee_{i\in\omega} \sigma_i$ if and only if $\mathcal{U}_i \leq \mathcal{V}$ for some i.

Our practice will be to express Maltsev conditions informally. For example, we will express the fact that variety $\mathcal{V}$ satisfies the strong Maltsev condition

$$\exists p\left(\mathrm{comp}^3_2(p, \pi^2_1, \pi^2_2, \pi^2_2) \approx \pi^2_1 \ \&\ \mathrm{comp}^3_2(p, \pi^2_2, \pi^2_2, \pi^2_1) \approx \pi^2_1\right), \tag{2.4}$$

which expresses the fact that $\mathcal{V}$ has a Maltsev term, by saying that $\mathcal{V}$ has a ternary term p such that the identities $p(x, y, y) \approx x$ and $p(y, y, x) \approx x$ hold in $\mathcal{V}$.

An n-ary element f of a clone $\mathcal{C}$ is **idempotent** if

$$\mathrm{comp}_1^n(f, \pi_1^1, \pi_1^1, \ldots, \pi_1^1) = \pi_1^1 .$$

If $\mathcal{C}$ is the clone of a variety $\mathcal{V}$, this means that $f(x, x, \ldots, x) \approx x$ is satisfied in $\mathcal{V}$. The idempotent elements of $\mathcal{C}$ form a subclone, $\mathrm{Id}(\mathcal{C})$. The **idempotent reduct** of $\mathcal{V}$ is the variety $\mathrm{Id}(\mathcal{V}) := \mathrm{Var}(\mathrm{Id}(\mathrm{Clo}(\mathcal{V})))$ that is associated to the idempotent subclone of $\mathrm{Clo}(\mathcal{V})$. A (strong) Maltsev condition is **idempotent** if for each term f in the condition the identity $f(x, x, \ldots, x) \approx x$ is a consequence of the identities of the Maltsev condition.

Two (strong) Maltsev conditions are **equivalent** if they define the same class of varieties.

LEMMA 2.11. *Any idempotent strong Maltsev condition is equivalent to one of the form $\exists F \bigwedge \Sigma$ where $F = \{h, k\}$, h is n-ary and k is n^2-ary, and Σ consists of the identities*

(i) $h(x, x, \ldots, x) \approx x$,
(ii) $k(x_{11}, \ldots, x_{nn}) \approx h\big(h(x_{11}, \ldots, x_{1n}), \ldots, h(x_{n1}, \ldots, x_{nn})\big)$, *plus*
(iii) *finitely many identities of the form* $k(\text{variables}) \approx k(\text{variables})$.

The proof of this lemma is part of the proof of Lemma 9.4 of [**34**].

A strong Maltsev condition $\exists F \bigwedge \Sigma$ is **linear** if each identity in Σ has the form $p(\text{variables}) \approx q(\text{variables})$ where $p, q \in F \cup \{\text{variables}\}$, while a Maltsev condition $\bigvee_{i \in \omega} \sigma_i$ is linear if each σ_i is.

LEMMA 2.12. *Any idempotent linear strong Maltsev condition is equivalent to one of the form $\exists F \bigwedge \Sigma$ where $F = \{f\}$, and Σ consists of the identities*

(i) $f(x, x, \ldots, x) \approx x$, *plus*
(ii) *finitely many identities of the form* $f(\text{variables}) \approx f(\text{variables})$.

This is proved by slightly modifying the proof in [**34**] of the preceding lemma.

All of the Maltsev conditions considered in this monograph are idempotent and linear, and will usually be expressible in the form described in Lemma 2.12 using only two variables. We introduce a notation for such 2-variable Maltsev conditions now.

Let f be an n-ary operation symbol and let $N = \{1, \ldots, n\}$. For each $U \subseteq N$, let $f_U(x, y)$ denote the term obtained from $f(x_1, \ldots, x_n)$ by substituting x for x_i if $i \in U$ and y for x_j if $j \notin U$. Observe that any identity of the form

$$f(\text{variables}) \approx f(\text{variables})$$

that uses only the variables x and y may be rewritten in the form $f_U(x, y) \approx f_V(x, y)$ where U is the set of positions where x appears on the left of the identity and V is the set of positions where x appears on the right. This particular identity may be abbreviated by $U \equiv V$. In order to encode a family $\mathcal{F}$ of 2-variable identities of the form $f(\text{variables}) \approx f(\text{variables})$, we define $\mathcal{B}(f)$ to be the Boolean algebra of subsets of N equipped with an equivalence relation E where $U \equiv_E V$ holds if and only if $f_U(x, y) \approx f_V(x, y)$ is a consequence of the identities in $\mathcal{F}$.

EXAMPLE 2.13. The statement that p is a Maltsev term for $\mathcal{V}$ may be expressed by saying that $\mathcal{V}$ satisfies identities of the form $p(\text{variables}) \approx p(\text{variables})$ if the idempotence of p is assumed, namely by $p(x, y, y) \approx p(x, x, x)$ and $p(y, y, x) \approx p(x, x, x)$. Here $N = \{1, 2, 3\}$ and $\mathcal{B}(p)$ is the Boolean algebra of Figure 2.2 with

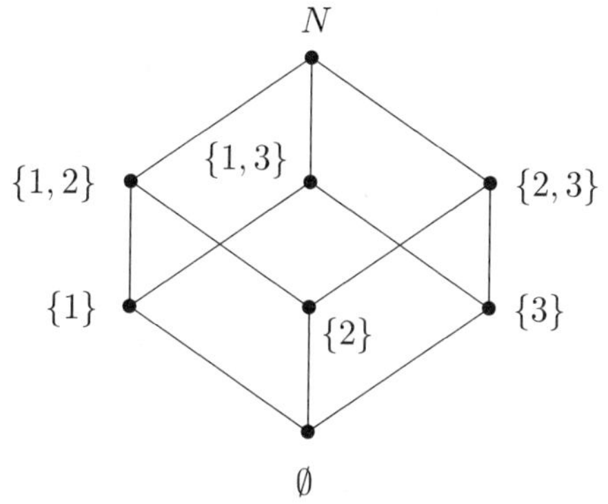

FIGURE 2.2. $\mathcal{B}(p)$

an equivalence relation E where $N \equiv_E \{1\} \equiv_E \{3\}$, since $p(x,x,x) \approx p(x,y,y) \approx p(y,y,x)$ are consequences of the starting identities, and $\emptyset \equiv_E \{1,2\} \equiv_E \{2,3\}$, since $p(y,y,y) \approx p(x,x,y) \approx p(y,x,x)$ are also consequences.

A subset of $\mathcal{B}(f)$ is **closed** if it is a union of E-classes. The next theorem is due to Walter Taylor.

THEOREM 2.14. *The following are equivalent for a variety $\mathcal{V}$.*

(1) *$\mathcal{V}$ satisfies a nontrivial idempotent Maltsev condition (i.e., one that fails in some variety).*
(2) *$\mathcal{V}$ satisfies an idempotent Maltsev condition that fails in the variety of sets.*
(3) *For some n, $\mathcal{V}$ has an idempotent n-ary term f such that $\mathcal{B}(f)$ has no closed ultrafilter.*

PROOF. See Corollary 5.3 of [**78**]. □

We will make frequent use of Theorem 2.14, so we introduce the following terminology.

DEFINITION 2.15. A term f is a **Taylor term** for a variety $\mathcal{V}$ if $\mathcal{V}$ satisfies $f(x,x,\dots,x) \approx x$ and enough other identities of the form $f_U(x,y) \approx f_V(x,y)$ so that $\mathcal{B}(f)$ has no closed ultrafilter.

Note that the word 'linear' does not appear in Conditions (1) and (2) of Theorem 2.14 although the Maltsev condition from (3) is linear. The fact that any variety satisfying a nontrivial idempotent Maltsev condition also satisfies one that is linear is the nontrivial part of the theorem.

Condition (3) of Theorem 2.14 means that if $\mathcal{U}$ is the ultrafilter of subsets of N containing the singleton $\{i\}$, then $\mathcal{U}$ is not closed. Therefore there is some $U \in \mathcal{U}$ and some $V \notin \mathcal{U}$ such that $\mathcal{V}$ satisfies $f_U(x,y) \approx f_V(x,y)$. Another way to express Condition (3) is that $\mathcal{V}$ satisfies an identity of the type $f(\text{variables}) \approx f(\text{variables})$ for each $1 \le i \le n$, where all variables are x or y, and x appears in the i-th position on the left while y appears in the i-th position on the right. An identity of this type will be called an **i-th Taylor identity** for f.

A result similar to Taylor's was discovered by David Hobby and Ralph McKenzie.

THEOREM 2.16. *The following are equivalent for a variety $\mathcal{V}$.*

(1) *$\mathcal{V}$ satisfies an idempotent Maltsev condition that fails in the variety of semilattices.*

(2) *For some n, $\mathcal{V}$ has an idempotent n-ary term f such that $\mathcal{B}(f)$ has no closed, proper, nonempty lattice filter.*

PROOF. See Lemma 9.5 of [**34**]. □

DEFINITION 2.17. A term f is a **Hobby–McKenzie term** for a variety $\mathcal{V}$ if $\mathcal{V}$ satisfies $f(x,x,\ldots,x) \approx x$ and enough other identities of the form $f_U(x,y) \approx f_V(x,y)$ so that $\mathcal{B}(f)$ has no closed, proper, nonempty lattice filter.

The term p from Example 2.13 is a Hobby–McKenzie term, and therefore also a Taylor term. To see this, suppose that $\mathcal{F}$ is a closed, proper, nonempty lattice filter of $\mathcal{B}(p)$. Then since $\mathcal{F}$ is a nonempty filter it contains the top element N. Since $\mathcal{F}$ is closed it must contain $\{1\}$ ($\equiv_E N$) and also $\{3\}$ ($\equiv_E N$). Since $\mathcal{F}$ is a lattice filter, it contains $\{1\} \cap \{3\} = \emptyset$. But any lattice filter containing $\emptyset$ is improper.

2.5. The Term Condition

Let $\mathbf{A} = \langle A; F\rangle$ be an algebra. An n-ary relation $R \subseteq A^n$ is **compatible** if it is a subalgebra of $\mathbf{A}^n$. If $\mathbf{B}$ is a subalgebra of $\mathbf{A}$, then the **restriction** of a relation $R \subseteq A^n$ to $\mathbf{B}$ is $R|_{\mathbf{B}} := R \cap B^n$. If δ is a congruence on $\mathbf{A}$, then $R/\delta := \{(a_1/\delta, \ldots, a_n/\delta) \mid (a_1,\ldots,a_n) \in R\}$. Both $R|_{\mathbf{B}}$ and R/δ are compatible if R is. If δ is a congruence, then R is **δ-closed** if $R = \delta \circ R \circ \delta$ (i.e., if $\mathbf{a}\ \delta\ \mathbf{b}\ R\ \mathbf{c}\ \delta\ \mathbf{d}$ implies $\mathbf{a}\ R\ \mathbf{d}$).

A compatible, reflexive, symmetric binary relation is called a **tolerance**. We will usually denote tolerances by upper case italic letters: $R, S, T, \ldots$. A compatible equivalence relation (i.e., a transitive tolerance) is a **congruence**, and congruences will usually be denoted by lower case Greek letters: α, β, $\gamma, \ldots$. The tolerance or congruence generated by a set $X \subseteq A \times A$ is usually denoted by $\mathrm{Tg}^{\mathbf{A}}(X)$ or $\mathrm{Cg}^{\mathbf{A}}(X)$ respectively, although if X contains only a few pairs then we may write, for example, $\mathrm{Cg}^{\mathbf{A}}(a,b)$ or $\mathrm{Cg}^{\mathbf{A}}\big((a,b),(c,d)\big)$ instead. If T is a tolerance on $\mathbf{A}$, then a maximal subset $B \subseteq A$ such that $B \times B \subseteq T$ is called a **block** of T. If T is in fact a congruence, then a block is the same thing as a congruence class. A tolerance or congruence is **trivial** if it is the equality relation and **nontrivial** otherwise.

The collection of congruences on $\mathbf{A}$, ordered by inclusion, is an algebraic lattice which is denoted $\mathbf{Con}(\mathbf{A})$. Its least and largest elements are denoted 0 and 1. Meet and join are denoted $\wedge$ and $\vee$ and are computed by $\alpha \wedge \beta = \alpha \cap \beta$ and $\alpha \vee \beta = \text{tr.cl.}(\alpha \cup \beta)$ where tr.cl. represents transitive closure.

An m-ary **polynomial operation** of $\mathbf{A}$ is an operation $f\colon A^m \to A$ such that $f(x_1,\ldots,x_m) = t^{\mathbf{A}}(x_1,\ldots,x_m,\mathbf{a})$ for some $(m+n)$-ary term t and some tuple $\mathbf{a} \in A^n$.

If S and T are tolerances on $\mathbf{A}$, then an **S,T-matrix** is a 2×2 matrix of elements of $\mathbf{A}$ of the form

$$\begin{bmatrix} p & q \\ r & s \end{bmatrix} = \begin{bmatrix} f(\mathbf{a},\mathbf{u}) & f(\mathbf{a},\mathbf{v}) \\ f(\mathbf{b},\mathbf{u}) & f(\mathbf{b},\mathbf{v}) \end{bmatrix}$$

where $f(\mathbf{x},\mathbf{y})$ is an $(m+n)$-ary polynomial of $\mathbf{A}$, $\mathbf{a}\ S\ \mathbf{b}$, and $\mathbf{u}\ T\ \mathbf{v}$. The set of all S,T-matrices is denoted $M(S,T)$.

Since tolerances are compatible with all polynomial operations, any two elements in the same row of an S,T-matrix are T-related and any two elements in the same column are S-related.

The fact that S and T are symmetric relations implies that $M(S,T)$ is closed under interchanging rows or columns:

$$\begin{bmatrix} p & q \\ r & s \end{bmatrix} \in M(S,T) \Leftrightarrow \begin{bmatrix} r & s \\ p & q \end{bmatrix} \in M(S,T) \Leftrightarrow \begin{bmatrix} s & r \\ q & p \end{bmatrix} \in M(S,T).$$

If $S = T$, then $M(S,T) = M(T,T)$ is also closed under transpose, as one sees by interchanging the roles of $\mathbf{x}$ and $\mathbf{y}$ in the polynomial $f(\mathbf{x}, \mathbf{y})$ that defines a given matrix.

DEFINITION 2.18. Let S and T be tolerances on an algebra $\mathbf{A}$, and let δ be a congruence on $\mathbf{A}$. If $p \equiv_\delta q$ implies that $r \equiv_\delta s$ whenever

$$\begin{bmatrix} p & q \\ r & s \end{bmatrix} \in M(S,T), \tag{2.5}$$

then we say that $\mathbf{C}(S,T;\delta)$ **holds**, or S **centralizes** T **modulo** δ.

By interchanging the rows of matrices one sees that $\mathbf{C}(S,T;\delta)$ holds if and only if

$$p \equiv_\delta q \quad \Longleftrightarrow \quad r \equiv_\delta s$$

for every S,T-matrix in (2.5).

The S,T-**term condition** is the condition $\mathbf{C}(S,T;0)$. There are other similar conditions called term conditions that we will meet later, but this is the original one.

When establishing that the implication defining $\mathbf{C}(S,T;\delta)$ holds, or when making use of the fact, we may use underlining to highlight places in equations or expressions where changes are to be made. For example, we may write the implication defining $\mathbf{C}(S,T;\delta)$ in the following form: If

$$f(\underline{\mathbf{a}}, \mathbf{u}) \equiv_\delta f(\underline{\mathbf{a}}, \mathbf{v}),$$

then

$$f(\underline{\mathbf{b}}, \mathbf{u}) \equiv_\delta f(\underline{\mathbf{b}}, \mathbf{v}).$$

The relation $\mathbf{C}(\ ,\ ;\ )$ is called the **centralizer relation**. The reason that this terminology is used is that when $\mathbf{A}$ is a group and S,T and δ are congruences on $\mathbf{A}$, then $\mathbf{C}(S,T;\delta)$ holds if and only if $[S,T] \leq \delta$ (see Chapter 1 of [**19**]).

The basic properties of the centralizer relation are enumerated in the following theorem.

THEOREM 2.19. *Let $\mathbf{A}$ be an algebra with tolerances S, S', T, T' and congruences α, α_i, β, δ, δ', δ_j. The following are true.*

(1) (Monotonicity in the first two variables) *If $\mathbf{C}(S,T;\delta)$ holds and $S' \subseteq S$, $T' \subseteq T$, then $\mathbf{C}(S',T';\delta)$ holds.*
(2) *$\mathbf{C}(S,T;\delta)$ holds if and only if $\mathbf{C}(\mathrm{Cg}^{\mathbf{A}}(S),T;\delta)$ holds.*
(3) *$\mathbf{C}(S,T;\delta)$ holds if and only if $\mathbf{C}(S,\delta \circ T \circ \delta;\delta)$ holds.*
(4) *If $T \cap \delta = T \cap \delta'$, then $\mathbf{C}(S,T;\delta) \iff \mathbf{C}(S,T;\delta')$.*
(5) (Semidistributivity in the first variable) *If $\mathbf{C}(\alpha_i,T;\delta)$ holds for all $i \in I$, then $\mathbf{C}(\bigvee_{i\in I}\alpha_i,T;\delta)$ holds.*
(6) *If $\mathbf{C}(S,T;\delta_j)$ holds for all $j \in J$, then $\mathbf{C}(S,T;\bigwedge_{j\in J}\delta_j)$ holds.*

(7) *If* $T \cap \big(S \circ (T \cap \delta) \circ S\big) \subseteq \delta$, *then* $\mathbf{C}(S, T; \delta)$ *holds.*
(8) *If* $\beta \wedge \big(\alpha \vee (\beta \wedge \delta)\big) \leq \delta$, *then* $\mathbf{C}(\alpha, \beta; \delta)$ *holds.*
(9) *Let* $\mathbf{B}$ *be a subalgebra of* $\mathbf{A}$. *If* $\mathbf{C}(S, T; \delta)$ *holds in* $\mathbf{A}$, *then* $\mathbf{C}(S|_{\mathbf{B}}, T|_{\mathbf{B}}; \delta|_{\mathbf{B}})$ *holds in* $\mathbf{B}$.
(10) *If* $\delta' \leq \delta$, *then the relation* $\mathbf{C}(S, T; \delta)$ *holds in* $\mathbf{A}$ *if and only if* $\mathbf{C}(S/\delta', T/\delta'; \delta/\delta')$ *holds in* $\mathbf{A}/\delta'$.

PROOF. Item (1) follows from the fact that $M(S', T') \subseteq M(S, T)$.

For (2), $\mathbf{C}\big(\mathrm{Cg}^{\mathbf{A}}(S), T; \delta\big) \implies \mathbf{C}(S, T; \delta)$ follows from (1), since $S \subseteq \mathrm{Cg}^{\mathbf{A}}(S)$. For the reverse implication (and also for the proof of item (5)), we will argue that if S_i is a tolerance, $\mathbf{C}(S_i, T; \delta)$ holds for all $i \in I$, and $\alpha :=$ tr.cl. $\big(\bigcup_{i \in I} S_i\big)$, then $\mathbf{C}(\alpha, T; \delta)$. (To complete the proof of (2) we need this only when $|I| = 1$, while in (5) we need it only when the S_i are congruences.)

Choose any matrix in $M(\alpha, T)$. If it is

$$\begin{bmatrix} p & q \\ r & s \end{bmatrix} = \begin{bmatrix} f(\mathbf{a}, \mathbf{u}) & f(\mathbf{a}, \mathbf{v}) \\ f(\mathbf{b}, \mathbf{u}) & f(\mathbf{b}, \mathbf{v}) \end{bmatrix},$$

then $\mathbf{a}$ is related to $\mathbf{b}$ by tr.cl. $\big(\bigcup_{i \in I} S_i\big)$. It is easy to see that there exist tuples $\mathbf{a} = \mathbf{a}_0 \; S_{i_1} \; \mathbf{a}_1 \; S_{i_2} \; \cdots \; S_{i_n} \; \mathbf{a}_n = \mathbf{b}$. These tuples determine matrices

$$\begin{bmatrix} p_k & q_k \\ p_{k+1} & q_{k+1} \end{bmatrix} := \begin{bmatrix} f(\mathbf{a}_k, \mathbf{u}) & f(\mathbf{a}_k, \mathbf{v}) \\ f(\mathbf{a}_{k+1}, \mathbf{u}) & f(\mathbf{a}_{k+1}, \mathbf{v}) \end{bmatrix} \in M(S_{i_{k+1}}, T).$$

We must show that $p \equiv_\delta q$ implies $r \equiv_\delta s$, so assume that $p \equiv_\delta q$. This is the same as $p_0 \equiv_\delta q_0$, and so by induction (using that $\mathbf{C}(S_{i_k}, T; \delta)$ holds for all k) we get that $p_k \equiv_\delta q_k$ for all k. Therefore $r = p_n \equiv_\delta q_n = s$. This completes the proofs of (2) and (5).

For (3), the implication $\mathbf{C}(S, \delta \circ T \circ \delta; \delta) \implies \mathbf{C}(S, T; \delta)$ follows from (1), since $T \subseteq \delta \circ T \circ \delta$. For the reverse implication, assume that $\mathbf{C}(S, T; \delta)$ holds, that

$$\begin{bmatrix} p & q \\ r & s \end{bmatrix} = \begin{bmatrix} f(\mathbf{a}, \mathbf{u}) & f(\mathbf{a}, \mathbf{v}) \\ f(\mathbf{b}, \mathbf{u}) & f(\mathbf{b}, \mathbf{v}) \end{bmatrix} \in M(S, \delta \circ T \circ \delta),$$

and that $p \equiv_\delta q$. There exist tuples $\mathbf{u}'$ and $\mathbf{v}'$ such that $\mathbf{u} \; \delta \; \mathbf{u}' \; T \; \mathbf{v}' \; \delta \; \mathbf{v}$. The matrix

$$\begin{bmatrix} p' & q' \\ r' & s' \end{bmatrix} = \begin{bmatrix} f(\mathbf{a}, \mathbf{u}') & f(\mathbf{a}, \mathbf{v}') \\ f(\mathbf{b}, \mathbf{u}') & f(\mathbf{b}, \mathbf{v}') \end{bmatrix}$$

is an S, T-matrix. Moreover,

$$p' = f(\mathbf{a}, \mathbf{u}') \; \delta \; f(\mathbf{a}, \mathbf{u}) = p \; \delta \; q = f(\mathbf{a}, \mathbf{v}) \; \delta \; f(\mathbf{a}, \mathbf{v}') = q'.$$

Since $\mathbf{C}(S, T; \delta)$ holds, it follows that $r' \equiv_\delta s'$. Hence

$$r = f(\mathbf{b}, \mathbf{u}) \; \delta \; f(\mathbf{b}, \mathbf{u}') = r' \; \delta \; s' = f(\mathbf{b}, \mathbf{v}') \; \delta \; f(\mathbf{b}, \mathbf{v}) = s,$$

or $r \equiv_\delta s$. This establishes $\mathbf{C}(S, \delta \circ T \circ \delta; \delta)$.

For (4), recall that elements in the same row of an S, T-matrix are T-related. So if $\begin{bmatrix} p & q \\ r & s \end{bmatrix} \in M(S, T)$, then since $T \cap \delta = T \cap \delta'$ we get that

$$p \equiv_\delta q \iff p \equiv_{T \cap \delta} q \iff p \equiv_{T \cap \delta'} q \iff p \equiv_{\delta'} q,$$

and

$$r \equiv_\delta s \iff r \equiv_{T \cap \delta} s \iff r \equiv_{T \cap \delta'} s \iff r \equiv_{\delta'} s.$$

Therefore the implication $p \equiv_\delta q \implies r \equiv_\delta s$ is equivalent to the implication $p \equiv_{\delta'} q \implies r \equiv_{\delta'} s$.

For (6), assume that $\begin{bmatrix} p & q \\ r & s \end{bmatrix} \in M(S,T)$. If $p \equiv q \pmod{\bigwedge \delta_j}$, then $p \equiv q \pmod{\delta_j}$ for all j. Since $\mathbf{C}(S,T;\delta_j)$ holds for all j we get that $r \equiv s \pmod{\delta_j}$ for all j, or equivalently that $r \equiv s \pmod{\bigwedge \delta_j}$. This shows that $\mathbf{C}(S,T;\bigwedge_{j\in J} \delta_j)$ holds.

For (7), choose an S,T-matrix $M = \begin{bmatrix} p & q \\ r & s \end{bmatrix}$. Assume that $p \equiv_\delta q$. Since the elements in the same row of M are T-related and the elements in the same column are S-related, we have $r\ S\ p\ T \cap \delta\ q\ S\ s$. Moreover, $r\ T\ s$ since these elements belong to the same row. Together this yields that $r\ T \cap \big(S \circ (T \cap \delta) \circ S\big)\ s$. By the assumption in (7), this implies that $r \equiv_\delta s$. This proves (7).

For item (8), if $\beta \wedge \big(\alpha \vee (\beta \wedge \delta)\big) \leq \delta$, then $\beta \cap \big(\alpha \circ (\beta \cap \delta) \circ \alpha\big) \leq \delta$, so $\mathbf{C}(\alpha,\beta;\delta)$ holds by (7).

Item (9) holds because any instance of the implication in Definition 2.18 defining $\mathbf{C}(S|_{\mathbf{B}}, T|_{\mathbf{B}}; \delta|_{\mathbf{B}})$ in $\mathbf{B}$ is an instance of the implication defining $\mathbf{C}(S,T;\delta)$ in $\mathbf{A}$.

For item (10), it suffices to observe that, when $\delta' \leq \delta$,

$$\begin{bmatrix} p'/\delta' & q'/\delta' \\ r'/\delta' & s'/\delta' \end{bmatrix} \in M(S/\delta', T/\delta')$$

if and only if there exist $p \equiv_{\delta'} p'$, $q \equiv_{\delta'} q'$, $r \equiv_{\delta'} r'$, and $s \equiv_{\delta'} s'$ with

$$\begin{bmatrix} p & q \\ r & s \end{bmatrix} \in M(S,T),$$

and that $p \equiv_\delta q \Leftrightarrow p'/\delta' \equiv_{\delta/\delta'} q'/\delta'$ and $r \equiv_\delta s \Leftrightarrow r'/\delta' \equiv_{\delta/\delta'} s'/\delta'$. □

Definition 2.20. The **commutator** of S and T, denoted by $[S,T]$, is the least congruence δ such that $\mathbf{C}(S,T;\delta)$ holds. T is **abelian** if $[T,T] = 0$. An algebra $\mathbf{A}$ is **abelian** if its largest congruence is.

By Theorem 2.19 (6), the class of all δ such that $\mathbf{C}(S,T;\delta)$ holds is closed under complete meet, so there is a least such δ. This implies that $[S,T]$ exists for any two tolerances S and T.

It is a well known fact, easily derivable from the definitions, that $\mathbf{A}$ is abelian if and only if the diagonal of $A \times A$ is a class of a congruence of $\mathbf{A} \times \mathbf{A}$.

Definition 2.21. The **centralizer of T modulo δ**, denoted by $(\delta : T)$, is the largest congruence α on $\mathbf{A}$ such that $\mathbf{C}(\alpha,T;\delta)$ holds.

By Theorem 2.19 (5), the class of all α such that $\mathbf{C}(\alpha,T;\delta)$ holds is closed under complete join, so there is a largest such α. This implies that $(\delta : T)$ exists for every δ and T. By Theorem 2.19 (2), the centralizer $(\delta : T)$ contains every tolerance S such that $\mathbf{C}(S,T;\delta)$ holds.

2.6. Congruence Identities

If $\mathcal{V}$ is a variety of algebras, then any lattice identity that holds in the class $\{\mathbf{Con}(\mathbf{A}) \mid \mathbf{A} \in \mathcal{V}\}$ of congruence lattices of algebras in $\mathcal{V}$ is called a **congruence identity** of $\mathcal{V}$. The **congruence variety** of $\mathcal{V}$, denoted CON($\mathcal{V}$), is the subvariety of $\mathcal{L}$ generated by $\{\mathbf{Con}(\mathbf{A}) \mid \mathbf{A} \in \mathcal{V}\}$, or alternatively is the variety of lattices axiomatized by the congruence identities that hold in $\mathcal{V}$. Similarly, a lattice quasi-identity that holds in congruence lattices of members of $\mathcal{V}$ is a **congruence quasi-identity** of $\mathcal{V}$.

The following theorem will be used in several places in this monograph.

THEOREM 2.22 (Cf. [**6**]). *Let Q be a quasi-identity satisfying (W). The class of varieties satisfying Q as a congruence quasi-identity is definable by a set of idempotent Maltsev conditions.*

Since congruence lattices are algebraic, and therefore meet continuous, it is reasonable to discuss **meet continuous congruence identities**, which we define to be meet continuous lattice identities considered as congruence identities. A meet continuous congruence identity is **trivial** if it holds in the congruence lattice of every algebra and **nontrivial** otherwise. It follows from Corollary 2.10 that free meet continuous lattices are algebraic, so they are isomorphic to congruence lattices according to the Grätzer–Schmidt Theorem [**26**]. Therefore a meet continuous lattice identity is nontrivial as a congruence identity if and only if it is nontrivial as a meet continuous lattice identity. Both statements mean the identity fails in some algebraic lattice.

It is proved in [**5**] that the class of varieties that satisfy the meet semidistributive law as a congruence quasi-identity is definable by a set of idempotent Maltsev conditions. An examination of that argument reveals that what is really proved is that, modulo the axioms defining $\mathcal{L}_{MC}$, the meet semidistributive law is equivalent to a meet continuous lattice identity.[1] In the next theorem, we prove that many different quasi-identities, including $\mathrm{SD}_\wedge$, are equivalent to meet continuous identities.

THEOREM 2.23. *Let p, q, r, s, $x_1, \ldots, x_n$ be lattice variables. Define $t := p \wedge (q \vee r)$. Let $w_i(p,q,r)$, $1 \le i \le m$, be ternary lattice words such that $w_i(p,q,r) \le q$ for some i, $w_j(p,q,r) \le r$ for some j, and $w_k(p,q,r) \le t$ for all k in the free lattice $\mathbf{F}_{\mathcal{L}}(\{p,q,r\})$. For each i let $x_i^* = (x_i \vee s) \wedge t$. For arbitrary n-ary lattice words u and v, the quasi-identity*

$$\bigwedge_{i=1}^{m} \big(w_i(p,q,r) \approx s\big) \to u(x_1^*, \ldots, x_n^*) \approx v(x_1^*, \ldots, x_n^*) \tag{2.6}$$

is equivalent to an identity in $\mathcal{L}_{MC}$.

PROOF. Let $f_i(x)$ denote the polynomial $w_i(p, q \vee x, r \vee x)$ of the free meet continuous lattice $\mathbf{F}_0 = \mathbf{F}_{\mathcal{L}_{MC}}(\{p,q,r\})$ (where $1 \le i \le m$). If $z \le t = p \wedge (q \vee r)$, then the idempotence of lattice words implies that for any i between 1 and m

$$\begin{aligned} \underline{z} = w_i(z,z,z) &\le w_i(p, q \vee z, r \vee z) = \underline{f_i(z)} \\ &\le w_i(p, q \vee t, r \vee t) \le w_i(p, q \vee r, q \vee r) \\ &\le p \wedge \big((q \vee r) \vee (q \vee r)\big) = \underline{t}\,. \end{aligned} \tag{2.7}$$

The underlined elements indicate why the principal ideal generated by $t = p \wedge (q \vee r)$ is closed under each f_i, and each f_i is an increasing function on this ideal. It follows from this that the set of all elements of the form $f_{i_1} \circ \cdots \circ f_{i_\ell}(0)$, $i_j \in \{1, \ldots, m\}$, is an updirected set of elements below t. The supremum of this set, $\delta^{\mathbf{w}}(p,q,r) \in F_0$, $\mathbf{w} := (w_1, \ldots, w_m)$, is therefore the least common fixed point in $\mathbf{F}_0$ of the polynomials f_i. Since $t = p \wedge (q \vee r)$ is a common fixed point of the f_i it follows that $\delta^{\mathbf{w}} \le t$ in $\mathbf{F}_0$. Now, if $\varphi\colon \mathbf{F}_0 \to \mathbf{L}$ is a surjective homomorphism of meet continuous lattices for which $a = \varphi(p)$, $b = \varphi(q)$ and $c = \varphi(r)$, then $\varphi\big(\delta^{\mathbf{w}}(p,q,r)\big) = \delta^{\mathbf{w}}(a,b,c)$ is the supremum of all $f^{\mathbf{L}}_{i_1} \circ \cdots \circ f^{\mathbf{L}}_{i_k}(0)$, where $f^{\mathbf{L}}_i(x) := w^{\mathbf{L}}_i(a, b \vee x, c \vee x)$. By an

[1] In fact, what is *really* shown is that the partial lattice $\mathbf{P}(\mathrm{SD}_\wedge)$ in the $\mathrm{SD}_\wedge$-configuration is projective relative to $\mathcal{L}_{MC}$.

argument like the one in (2.7), each $f_i^{\mathbf{L}}$ is increasing on the principal ideal generated by $t^{\mathbf{L}}(a,b,c) = a \wedge (b \vee c)$ and maps this ideal into itself. Hence $\delta^{\mathbf{w}}(a,b,c)$ is the least common fixed point of the polynomials $f_i^{\mathbf{L}}$ in $\mathbf{L}$.

For each variable x_i, set $x_i^+ := \big(x_i \vee \delta^{\mathbf{w}}(p,q,r)\big) \wedge t$, which is an element of $\mathbf{F} := \mathbf{F}_{\mathcal{L}_{MC}}(\{p,q,r,s,x_1,\dots,x_n\})$. We are now prepared to argue that quasi-identity (2.6) is equivalent to the identity

$$u(x_1^+,\dots,x_n^+) \approx v(x_1^+,\dots,x_n^+) \tag{2.8}$$

in $\mathcal{L}_{MC}$.

Assume first that $\mathbf{L}$ is a meet continuous lattice satisfying (2.8). To show that $\mathbf{L}$ satisfies (2.6), choose any assignment of the variables $\{p,q,r,s,\mathbf{x}\} \to L$ that satisfies the premises of quasi-identity (2.6), and let $\psi\colon \mathbf{F} \to \mathbf{L}$ be the extension of this assignment to a homomorphism. Equivalently, choose a homomorphism ψ so that if

$$\big(\psi(p),\psi(q),\psi(r),\psi(s),\psi(\mathbf{x})\big) = (a,b,c,d,\mathbf{g}),$$

then $w_i^{\mathbf{L}}(a,b,c) = d$ for all i. Define $e := \psi(t) = a \wedge (b \vee c)$ and $g_i^* := \psi(x_i^*) = (g_i \vee d) \wedge e$. We must show that $u^{\mathbf{L}}(\mathbf{g}^*) = v^{\mathbf{L}}(\mathbf{g}^*)$.

Since $w_i(p,q,r) \le q$ for some i, $w_j(p,q,r) \le r$ for some j, and $w_k(p,q,r) \le \big(p \wedge (q \vee r)\big) \le p$ for all k in $\mathbf{F}_0$, it follows that

$$d = w_i(a,b,c) = w_j(a,b,c) = w_k(a,b,c) \le a \wedge b \wedge c$$

in $\mathbf{L}$. Since $d = w_1(a,b,c) \ge a \wedge b \wedge c$ we even have $d = a \wedge b \wedge c$. Therefore $f_i^{\mathbf{L}}(d) = w_i^{\mathbf{L}}(a, b \vee d, c \vee d) = w_i^{\mathbf{L}}(a,b,c) = d$ for all i, yielding that d is a common fixed point of the f_i in $\mathbf{L}$. Since the meet continuous sublattice $\mathbf{L}_0 \le \mathbf{L}$ that is generated by a, b and c contains d and has $\delta^{\mathbf{w}}(a,b,c)$ as its least common fixed point of the f_i's, it follows that $\delta^{\mathbf{w}}(a,b,c) \le d$. Since $d = a \wedge b \wedge c \le f_1(0) \le \delta^{\mathbf{w}}(a,b,c)$, we even have $d = \delta^{\mathbf{w}}(a,b,c)$. This implies that $\psi(x_i^+) = \psi(x_i^*) = g_i^*$. This gives us the desired result

$$u^{\mathbf{L}}(\mathbf{g}^*) = \psi\big(u(\mathbf{x}^+)\big) = \psi\big(v(\mathbf{x}^+)\big) = v^{\mathbf{L}}(\mathbf{g}^*),$$

where the middle equality is from (2.8).

Now suppose conversely that $\mathbf{L}$ satisfies (2.6). To verify that $\mathbf{L}$ satisfies (2.8), choose any variable assignment in $\mathbf{L}$ and extend it to a homomorphism $\psi\colon \mathbf{F} \to \mathbf{L}$. We must show that $\psi\big(u(\mathbf{x}^+)\big) = \psi\big(v(\mathbf{x}^+)\big)$. Since the variable s does not appear in (2.8) we need only consider homomorphisms ψ for which $\psi(s) = \psi\big(\delta^{\mathbf{w}}(p,q,r)\big)$, i.e., those of the form

$$\big(\psi(p),\psi(q),\psi(r),\psi(s),\psi(\mathbf{x})\big) = (a,b,c,\delta^{\mathbf{w}}(a,b,c),\mathbf{g}).$$

Under such homomorphisms

$$\psi(x_i^+) = \psi(x_i^*) = \big(g_i \vee \delta^{\mathbf{w}}(a,b,c)\big) \wedge \big(a \wedge (b \vee c)\big).$$

The element $\delta := \delta^{\mathbf{w}}(a,b,c)$ is the least common fixed point of the polynomials $f_i^{\mathbf{L}}(x) := w^{\mathbf{L}}(a, b \vee x, c \vee x)$ in the meet continuous sublattice of $\mathbf{L}$ generated by a, b and c. Thus, if $b^\delta := b \vee \delta$ and $c^\delta := c \vee \delta$, we get that

$$\delta = f_i^{\mathbf{L}}(\delta) = w_i(a, b \vee \delta, c \vee \delta) = w_i(a, b^\delta, c^\delta)$$

for all i. Thus, each premise $w_i(a,b^\delta,c^\delta) = \delta$ of quasi-identity (2.6) is satisfied by the variable assignment $\psi^\delta\colon (p,q,r,s,\mathbf{x}) \mapsto (a,b^\delta,c^\delta,\delta,\mathbf{g})$. Since $\mathbf{L}$ satisfies (2.6), we get that

$$u^{\mathbf{L}}\big(\psi^{\delta}(\mathbf{x}^*)\big) = v^{\mathbf{L}}\big(\psi^{\delta}(\mathbf{x}^*)\big) \tag{2.9}$$

holds in $\mathbf{L}$, where

$$\psi^{\delta}(x_i^*) = (g_i \vee \delta) \wedge \big(a \wedge (b^{\delta} \vee c^{\delta})\big)\,.$$

But $\delta^{\mathbf{w}}(p,q,r) \leq p \wedge (q \vee r)$ in $\mathbf{F}_{\mathcal{L}_{MC}}(\{p,q,r\})$, so

$$\delta = \delta^{\mathbf{w}}(a,b,c) \leq a \wedge (b \vee c)$$

in $\mathbf{L}$. This yields the last equality in

$$a \wedge (b \vee c) \leq a \wedge (b^{\delta} \vee c^{\delta}) = a \wedge (b \vee c \vee \delta) = a \wedge (b \vee c)\,.$$

This shows that $a \wedge (b^{\delta} \vee c^{\delta}) = a \wedge (b \vee c)$, so

$$\psi^{\delta}(x_i^*) = (g_i \vee \delta) \wedge \big(a \wedge (b \vee c)\big) = \psi(x_i^+)\,.$$

From this and (2.9) we derive that

$$\psi\big(u(\mathbf{x}^+)\big) = u^{\mathbf{L}}\big(\psi^{\delta}(\mathbf{x}^*)\big) = v^{\mathbf{L}}\big(\psi^{\delta}(\mathbf{x}^*)\big) = \psi\big(v(\mathbf{x}^+)\big)\,,$$

as desired. This proves that (2.8) holds in $\mathbf{L}$. □

REMARK 2.24. If $\mathbf{L}$ is a meet continuous lattice, then for every homomorphism $\psi\colon \mathbf{F} \to \mathbf{L}$ for which $(p,q,r,s,\mathbf{x}) \mapsto (a,b,c,d,\mathbf{g})$ and $w_i(a,b,c) = d$ for all i, it must be that d is below $\psi(t) = a \wedge (b \vee c) =: e$, since $w_i(p,q,r) \leq t$. Call $I := I[d,e]$ a "bad interval" in $\mathbf{L}$ if it arises in this way from elements satisfying the premises of quasi-identity (2.6). For each i we have $\psi(x_i^*) = (g_i \vee d) \wedge e \in I$, and conversely for any $h \in I$ we have $h = \psi(x_i^*)$ if ψ assigns h to x_i. Thus, $\psi(x_i^*)$ may be viewed as a typical element of I. Altogether this means that the premises of (2.6) identify which intervals are bad, and the quasi-identity itself asserts that all bad intervals satisfy the identity $u \approx v$.

If we choose $m = n = 2$, $w_1(p,q,r) = p \wedge q$, $w_2(p,q,r) = p \wedge r$, then the bad intervals are the $\mathrm{SD}_{\wedge}$-failures. If $u(x_1,x_2) = x_1$ and $v(x_1,x_2) = x_2$, then (2.6) asserts precisely that all $\mathrm{SD}_{\wedge}$-failures satisfy $x_1 \approx x_2$, i.e., they are trivial. These choices for w_i, u and v show that the meet semidistributive law is equivalent to a quasi-identity of type (2.6), hence to a meet continuous identity.

Consider choosing the premises of (2.6) in the same way as in the previous paragraph, but allowing $u \approx v$ to be some other lattice identity. We obtain from Theorem 2.23 that the class of meet continuous lattices whose $\mathrm{SD}_{\wedge}$-failures satisfy $u \approx v$ is a subvariety of $\mathcal{L}_{MC}$.

Now consider modifying the choice of the premises of (2.6) from our earlier choice, but keeping $u(x_1,x_2) = x_1$ and $v(x_1,x_2) = x_2$. In this case, (2.6) still asserts that the bad intervals are trivial, but the bad intervals need not be the $\mathrm{SD}_{\wedge}$-failures. Instead, the bad intervals are those of the form $I[d,e]$ where for some $a,b,c \in L$ we have $w_i(a,b,c) = d$ for all i and $e = a \wedge (b \vee c)$. A more direct way of asserting that $I[d,e]$ is trivial is to assert that $e = d$ (where $e = a \wedge (b \vee c)$). This is achieved by rewriting quasi-identity (2.6) in the form

$$\bigwedge_{i=1}^{m} \big(w_i(p,q,r) \approx s\big) \to \big((p \wedge (q \vee r)) = s\big)\,. \tag{2.10}$$

Quasi-identity (2.10) is equivalent modulo the identities defining lattice theory to quasi-identity (2.6) when $u(x_1, x_2) = x_1$ and $v(x_1, x_2) = x_2$. In particular, Theorem 2.23 proves that a quasi-identity obtained from the meet semidistributive law,

$$\big((p \wedge q) \approx s\big) \ \&\ \big((p \wedge r) \approx s\big) \to \big((p \wedge (q \vee r)) = s\big)\,,$$

by adding additional premises $w_i(p, q, r) \approx s$ (where $w_i(p, q, r) \leq t$) is equivalent to a meet continuous identity.

By examining the proof of Theorem 2.23 one finds that the bad intervals in $\mathbf{L}$ associated to a quasi-identity of type (2.6) or (2.10) are those of the form $I[d, e]$ where for some $a, b, c \in L$ it is the case that $d = \delta^{\mathbf{w}}(a, b, c)$ and $e = a \wedge (b \vee c)$. Since quasi-identity (2.10) asserts only that bad intervals are trivial, a simpler meet continuous identity equivalent to (2.10) is

$$\delta^{\mathbf{w}}(p, q, r) \approx \big(p \wedge (q \vee r)\big)\,. \tag{2.11}$$

It is not hard to describe explicitly the meet continuous identity that is equivalent to the meet semidistributive law. Define lattice words in the variables p, q, r by $q_0 = q$, $r_0 = r$, $q_{n+1} = q \vee (p \wedge r_n)$, and $r_{n+1} = r \vee (p \wedge q_n)$. Let $q_\omega = \bigvee_{n<\omega} q_n$ and let $r_\omega = \bigvee_{n<\omega} r_n$.

COROLLARY 2.25. *A meet continuous lattice is meet semidistributive if and only if it satisfies the identity*

$$p \wedge (q \vee r) \approx p \wedge q_\omega\,. \tag{2.12}$$

PROOF. This is an instance of the comments in the final paragraph of Remark 2.24 above. For the choices $w_1(p, q, r) = p \wedge q$ and $w_2(p, q, r) = p \wedge r$ the polynomials f_i that occur in the proof of Theorem 2.23 are $f_1(x) = p \wedge (q \vee x) =: g(x)$ and $f_2(x) = p \wedge (r \vee x) =: h(x)$. Since these are idempotent polynomials of $\mathbf{F}_{\mathcal{L}_{MC}}(p, q, r)$, the words of the form $f_{i_1} \circ \cdots \circ f_{i_\ell}(0)$ that occur in the preceding proof reduce to alternating compositions $(gh)^k(0)$, $(hg)^k(0)$, $g(hg)^k(0)$, or $h(gh)^k(0)$. The join of these words is $\delta^{\mathbf{w}}(p, q, r) = \bigvee_{k<\omega}(gh)^k(0) = \bigvee_{k<\omega}(hg)^k(0)$. But $g(0) = p \wedge q_0$, $h(0) = p \wedge r_0$, $g(p \wedge r_n) = p \wedge q_{n+1}$ and $h(p \wedge q_n) = p \wedge r_{n+1}$, so $\delta^{\mathbf{w}}(p, q, r) = \bigvee_{k \text{ odd}} p \wedge q_k = p \wedge q_\omega$ by meet continuity, and also $\delta^{\mathbf{w}}(p, q, r) = \bigvee_{k \text{ even}} p \wedge r_k = p \wedge r_\omega$. Hence the meet semidistributive law is equivalent to $p \wedge q_\omega \approx p \wedge (q \vee r)$ or to $p \wedge r_\omega \approx p \wedge (q \vee r)$ for meet continuous lattices. □

This proof shows that $p \wedge q_\omega \approx p \wedge r_\omega$, so identity (2.12) can be written as a weakened distributive law:

$$p \wedge (q \vee r) \approx (p \wedge q_\omega) \vee (p \wedge r_\omega)\,.$$

CHAPTER 3

Strong Term Conditions

In this chapter we introduce the term conditions that define strongly abelian and strongly rectangular congruences. Our purpose is to show that a variety satisfies a nontrivial idempotent Maltsev condition if and only if it has no member with a nonzero strongly abelian congruence, or equivalently has no member with a nonzero strongly rectangular congruence.

3.1. Varieties Omitting Strongly Abelian Congruences

Ralph McKenzie introduced the notion of "strong abelianness" in [**67**], and, with David Hobby, showed in [**34**] that it is a natural and useful concept in the study of finite algebras. McKenzie also discovered a quite different abelianness concept, called "rectangulation" in [**55**], which made sense for finite subdirectly irreducible algebras with nonabelian monolith, and which has applications to the study of residually small varieties.

In [**49**], we required a commutator theory associated to the strong abelianness concept. In the process of developing one, we discovered a new natural and useful type of abelianness resembling McKenzie's rectangulation concept that we will call in this monograph "strong rectangulation". In [**49**] we developed a commutator theory for strong rectangulation, too. In this chapter we explore these two theories further. In Chapter 5 we will introduce a new definition for McKenzie's original rectangulation concept that makes sense for any algebra, and initiate the development of a commutator theory for it.

It is important to draw attention to the fact that we used the term "rectangulation" in [**49**] for what will henceforth be called "strong rectangulation".

DEFINITION 3.1. Let S and T be tolerances on an algebra $\mathbf{A}$, and let δ be a congruence on $\mathbf{A}$. If $q \equiv_\delta r$ implies that $r \equiv_\delta s$ whenever

$$\begin{bmatrix} p & q \\ r & s \end{bmatrix} \in M(S,T), \tag{3.1}$$

then we say that $\mathbf{SR}(S,T;\delta)$ **holds**, or S **strongly rectangulates** T **modulo** δ. We say that $\mathbf{S}(S,T;\delta)$ **holds**, or S **strongly centralizes** T **modulo** δ if $\mathbf{SR}(S,T;\delta)$ and $\mathbf{C}(S,T;\delta)$ both hold.

A tolerance T is **strongly rectangular** if $\mathbf{SR}(T,T;0)$ holds, and is **strongly abelian** if $\mathbf{S}(T,T;0)$ holds.

By interchanging the rows and columns of matrices one sees that $\mathbf{SR}(S,T;\delta)$ holds if and only if

$$q \equiv_\delta r \iff p \equiv_\delta q \equiv_\delta r \equiv_\delta s \iff p \equiv_\delta s$$

for every S,T-matrix in (3.1). It follows from this that $\mathbf{SR}(S,T;\delta)$ is equivalent to $\mathbf{SR}(T,S;\delta)$.

Now, and later, we must consider the issue of whether nonobvious relationships hold among the term conditions that we introduce. For example, is every abelian tolerance strongly abelian?[1] There is an all-purpose example for answering this type of question, which may be used to demonstrate the independence of any combination of the term conditions that we define in this monograph.

EXAMPLE 3.2. Let X be a set and let $\mathcal{M} \subseteq X^{2\times 2}$ be a set of 2×2 matrices with entries in X. For $M \in \mathcal{M}$, let $Y_M = \{a_M, b_M, u_M, v_M\}$. Let $Y = \bigcup_{M\in\mathcal{M}} Y_M$. Let A be the disjoint union $X \cup Y \cup \{0\}$. For each $M = \begin{bmatrix} p & q \\ r & s \end{bmatrix} \in \mathcal{M}$ define a binary operation f_M on A by

$$f_M(x, y) := \begin{cases} p & \text{if } (x, y) = (a_M, u_M); \\ q & \text{if } (x, y) = (a_M, v_M); \\ r & \text{if } (x, y) = (b_M, u_M); \\ s & \text{if } (x, y) = (b_M, v_M); \text{ and} \\ 0 & \text{otherwise.} \end{cases}$$

Let $\mathbf{A}(\mathcal{M}) = \langle A; \{f_M \mid M \in \mathcal{M}\}\rangle$, $R = \mathrm{Tg}^{\mathbf{A}(\mathcal{M})}(\{(a_M, b_M) \mid M \in \mathcal{M}\})$, $S = \mathrm{Tg}^{\mathbf{A}(\mathcal{M})}(\{(u_M, v_M) \mid M \in \mathcal{M}\})$, and $T = \mathrm{Tg}^{\mathbf{A}(\mathcal{M})}(R \cup S)$.

CLAIM 3.3. *The following hold.*

(1) *The set $M(R, S)$ consists only of the trivial R, S-matrices, namely those of the form*

$$\begin{bmatrix} p & p \\ q & q \end{bmatrix} \quad \text{and} \quad \begin{bmatrix} r & s \\ r & s \end{bmatrix}$$

with $(p, q) \in R$ and $(r, s) \in S$, together with the closure of $\mathcal{M}$ under interchanging rows and columns.

(2) *$M(T, T)$ consists only of the trivial T, T-matrices together with the closure of $\mathcal{M}$ under interchanging rows and columns and under transpose.*

We leave the proof of this claim as an exercise, guided by the following comments.

(i) The trivial matrices come from using a projection operation as the term in the matrix.

(ii) Any nontrivial composition of f_M's is zero, so only $f_M(x, y)$ and $f_M(y, x)$ can be used to create a nontrivial matrix. However $f_M(y, x)$ does not produce any nontrivial R, S-matrices.

We also leave the verification of the following as an exercise: the claim remains true if you replace R, S and T by the congruences they generate. And, as a final exercise: any failure of any term condition introduced in this monograph must involve a nontrivial R, S or T, T-matrix.

What this means is that when considering relationships between term conditions we may ignore any algebraic considerations and think completely set-theoretically. If it is possible to write down matrices whose patterns of entries formally conflict with the satisfaction of one term condition but do not formally conflict with the satisfaction of another, then it is possible to build an algebra that has matrices with these properties (subject to the restrictions of (1) and (2) of Claim 3.3).

[1] No.

For example, to show that there exists an algebra that has an abelian tolerance that is not strongly abelian, it is enough to observe that any T, T-matrix of the form $M = \begin{bmatrix} p & q \\ r & s \end{bmatrix} = \begin{bmatrix} 0 & 1 \\ 1 & 0 \end{bmatrix}$ formally conflicts with the satisfaction of $\mathbf{SR}(T, T; 0)$ (since $q = r \neq s$), but does not formally conflict with $\mathbf{C}(T, T; 0)$ (even after interchanging the rows or columns, or taking the transpose). This suggests that we take $X = \{0, 1\}$ and $\mathcal{M} = \{M\}$. By Claim 3.3, the nontrivial T, T-matrices of $\mathbf{A}(\mathcal{M})$ are

$$\begin{bmatrix} 0 & 1 \\ 1 & 0 \end{bmatrix} \quad \text{and} \quad \begin{bmatrix} 1 & 0 \\ 0 & 1 \end{bmatrix}.$$

This implies that the tolerance T on $\mathbf{A}(\mathcal{M})$ is abelian but not strongly rectangular (hence not strongly abelian).

In Lemma 2.2 of [**49**] we state and prove a result for $\mathbf{SR}(\ ,\ ;\ )$ similar to Theorem 2.19 for $\mathbf{C}(\ ,\ ;\ )$. These two results can be combined to produce a similar result for $\mathbf{S}(\ ,\ ;\ )$. Here we describe only the results about the strong centrality relation and the strong rectangulation relation that will be used in this chapter.

THEOREM 3.4. *Let $\mathbf{A}$ be an algebra with tolerances S, S', T, T' and congruences δ, δ', δ_j. Let $\mathbf{Q} = \mathbf{S}$ or $\mathbf{SR}$. The following are true.*

(1) *If $\mathbf{Q}(S, T; \delta)$ holds and $S' \subseteq S$, $T' \subseteq T$, then $\mathbf{Q}(S', T'; \delta)$ holds.*
(2) *If $\mathbf{Q}(S, T; \delta_j)$ holds for all $j \in J$, then $\mathbf{Q}(S, T; \bigwedge_{j \in J} \delta_j)$ holds.*
(3) *Let $\mathbf{B}$ be a subalgebra of $\mathbf{A}$. If $\mathbf{Q}(S, T; \delta)$ holds in $\mathbf{A}$, then $\mathbf{Q}(S|_{\mathbf{B}}, T|_{\mathbf{B}}; \delta|_{\mathbf{B}})$ holds in $\mathbf{B}$.*
(4) *If $\delta' \leq \delta$, then the relation $\mathbf{Q}(S, T; \delta)$ holds in $\mathbf{A}$ if and only if $\mathbf{Q}(S/\delta', T/\delta'; \delta/\delta')$ holds in $\mathbf{A}/\delta'$.*
(5) *If $T \cap \big(((S \circ T) \cap \delta) \circ S\big) \subseteq \delta$, then $\mathbf{SR}(S, T; \delta)$ holds. If $T \cap \big(S \circ (T \cap \delta) \circ S\big) \subseteq \delta$ also holds, then $\mathbf{S}(S, T; \delta)$ also holds.*

PROOF. Items (1)–(4) are proved in exactly the same way as items (1), (6), (9) and (10) of Theorem 2.19.

For item (5), choose any S, T-matrix $M = \begin{bmatrix} p & q \\ r & s \end{bmatrix}$. Assume that $r \equiv_\delta q$. Since the elements in the same row of M are T-related and the elements in the same column are S-related, we have $r\ S\ p\ T\ q$ and $r\ \delta\ q$, so $r\ (S \circ T) \cap \delta\ q$. Hence $r\ (S \circ T) \cap \delta\ q\ S\ s$ and $r\ T\ s$ imply $(r, s) \in T \cap \big(((S \circ T) \cap \delta) \circ S\big) \subseteq \delta$. Since M was chosen arbitrarily, $\mathbf{SR}(S, T; \delta)$ holds. If now $T \cap \big(S \circ (T \cap \delta) \circ S\big) \subseteq \delta$ also holds, then $\mathbf{C}(S, T; \delta)$ also holds by Theorem 2.19 (7). Hence $\mathbf{S}(S, T; \delta)$ also holds. □

Theorem 3.4 (2) implies that for $\mathbf{Q} = \mathbf{S}$ or $\mathbf{SR}$ there is a least δ such that $\mathbf{Q}(S, T; \delta)$ holds, and we denote the least such δ by $[S, T]_{\mathbf{Q}}$.

LEMMA 3.5. *Let $\mathcal{V}$ be a variety, $\mathbf{F} = \mathbf{F}_{\mathcal{V}}(x, y)$ the free $\mathcal{V}$-algebra over $X = \{x, y\}$, and $\theta = \mathrm{Cg}^{\mathbf{F}}(x, y)$. If $\mathbf{Q} = \mathbf{S}$ or $\mathbf{SR}$, then the following are equivalent.*

(1) *There exists an algebra $\mathbf{A} \in \mathcal{V}$ that has a nonzero $\mathbf{Q}$-abelian congruence (i.e., a congruence α such that $[\alpha, \alpha]_{\mathbf{Q}} = 0$).*
(2) $(x, y) \notin [\theta, \theta]_{\mathbf{Q}}$.

Now let T be the tolerance on $\mathbf{F}$ generated by the pair (x, y). The following are equivalent.

(3) *There exists an algebra* $\mathbf{A} \in \mathcal{V}$ *that has a nonzero* $\mathbf{Q}$*-abelian tolerance (i.e., a tolerance* S *such that* $[S,S]_{\mathbf{Q}} = 0$*).*
(4) $(x,y) \notin [T,T]_{\mathbf{Q}}$.

PROOF. Both halves of the lemma are proved in the same way, so we prove the first half only.

Assume that α is a nonzero $\mathbf{Q}$-abelian congruence on $\mathbf{A}$. Choose $(a,b) \in \alpha$ with $a \neq b$. Then from Theorem 3.4 (3) and (1) we get that if $\mathbf{B}$ is the subalgebra of $\mathbf{A}$ generated by $\{a,b\}$ and $\beta = \mathrm{Cg}^{\mathbf{B}}(a,b)$, then β is $\mathbf{Q}$-abelian. Let $h\colon \mathbf{F} \to \mathbf{B}$ be the homomorphism sending x to a and y to b. This homomorphism is surjective since $\mathbf{B}$ is generated by a and b. Let $\kappa = \ker(h)$ and $\lambda = h^{-1}(\beta)$. As $\big(h(x), h(y)\big) = (a,b) \in \beta \setminus 0$ we have $(x,y) \in \lambda \setminus \kappa$, which means that $\theta \leq \lambda$ and $\theta \not\leq \kappa$. From Theorem 3.4 (4) we know that $\mathbf{Q}(\lambda,\lambda;\kappa)$ holds in $\mathbf{F}$ if and only if $\mathbf{Q}(\lambda/\kappa, \lambda/\kappa; 0)$ holds in $\mathbf{F}/\kappa$. But it does, since λ/κ corresponds to β under the natural isomorphism of $\mathbf{F}/\kappa$ with $\mathbf{B}$. Therefore, by Theorem 3.4 (1), we derive that $\mathbf{Q}(\theta,\theta;\kappa)$ holds. We have $\mathbf{Q}(\theta,\theta;\theta)$ by Theorem 3.4 (5), and therefore $\mathbf{Q}(\theta,\theta;\kappa\wedge\theta)$ by Theorem 3.4 (2). This proves that $[\theta,\theta]_{\mathbf{Q}} \leq \kappa \wedge \theta < \theta$, so $(x,y) \notin [\theta,\theta]_{\mathbf{Q}}$.

On the other hand, if $(x,y) \notin [\theta,\theta]_{\mathbf{Q}}$, then Theorem 3.4 (4) guarantees that $\theta/[\theta,\theta]_{\mathbf{Q}}$ is a $\mathbf{Q}$-abelian congruence on $\mathbf{F}/[\theta,\theta]_{\mathbf{Q}}$ that relates the distinct elements $x/[\theta,\theta]_{\mathbf{Q}}$ and $y/[\theta,\theta]_{\mathbf{Q}}$. This shows that $\mathcal{V}$ contains an algebra with a nonzero $\mathbf{Q}$-abelian congruence. □

Our goal in this section is to prove that the four conditions of Lemma 3.5 are equivalent to each other, and that when $\mathbf{Q} = \mathbf{S}$ the conditions hold for a variety if and only if the variety has no Taylor term (Definition 2.15). This is also true when $\mathbf{Q} = \mathbf{SR}$, and that will be proved in the next section. To prove the equivalence of (1) and (2) with (3) and (4) we start by analyzing the relationship between the congruence θ and the tolerance T from Lemma 3.5. This is done in a little more generality than is immediately necessary, in that the next lemma considers free algebras on more than two variables, but this will allow us to avoid the need to generalize the result later.

LEMMA 3.6. *Let* $\mathcal{V}$ *be any variety and let* $\mathbf{F} = \mathbf{F}_{\mathcal{V}}(X)$ *be the free* $\mathcal{V}$*-algebra generated by* X*. Let* E *be an equivalence relation on* X*, and* $\tau\colon X \to X$ *a transversal for* E*. That is,* τ *is a function from* X *to* X *satisfying*

(i) $x \mathrel{E} \tau(x)$*, and*
(ii) $x \mathrel{E} y \implies \tau(x) = \tau(y)$.

Let $\theta = \mathrm{Cg}^{\mathbf{F}}(E)$ *and* $T = \mathrm{Tg}^{\mathbf{F}}(E)$*. The following hold.*

(1) θ *is the kernel of the endomorphism* $h\colon \mathbf{F} \to \mathbf{F}$ *induced by* $\tau\colon X \to X$*.*
(2) $\big(a, h(a)\big) \in T$ *for any* $a \in F$*.*
(3) $T \circ T = \theta$*.*
(4) *If* $B := \{u(\mathbf{x}) \in F \mid u \text{ is an idempotent term}\}$*,* $E = X \times X$*, and* $\tau(X) = \{x\}$*, then* $B = x/\theta$ *and* B *is the unique* T*-block containing* x*.*

PROOF. Observe that the fact that τ is a transversal for E implies that $\tau\circ\tau = \tau$ and $E = \ker(\tau)$.

We prove (2) first. If $a \in F$, then $a = t(x_1, \ldots, x_n)$ for some term t. Since T is a compatible relation containing $\big(x_i, \tau(x_i)\big)$ for every i, it also contains $t\big((x_1, \tau(x_1)), \ldots, (x_n, \tau(x_n))\big) = \big(t(\mathbf{x}), h(t(\mathbf{x}))\big) = \big(a, h(a)\big)$.

For (1), the fact that h is induced by an idempotent function on generators implies that h is an idempotent endomorphism, therefore h is a transversal for $\ker(h)$. Since $\ker(\tau) \subseteq \ker(h)$ we have $\theta \subseteq \ker(h)$. It follows that every θ-block contains at most one element of $\mathrm{im}(h)$. On the other hand, every θ-block contains at least one element of $\mathrm{im}(h)$, since if a is in a θ-block B, then $h(a)$ is certainly in $\mathrm{im}(h)$, while $(a, h(a)) \in T \subseteq \theta$ by (2). Thus $h(a) \in \mathrm{im}(h) \cap B$. This shows that h is also a transversal for θ. Since comparable equivalence relations with a common transversal are equal, $\theta = \ker(h)$.

For (3), $T \circ T \subseteq \mathrm{Cg}^{\mathbf{F}}(T) = \theta$, so it is enough to show that $T \circ T \supseteq \theta$. If $(a, b) \in \theta$, then $h(a) = h(b)$. By (2), we have

$$a\; T\; h(a) = h(b)\; T\; b\,,$$

proving that $(a, b) \in T \circ T$, indeed.

For (4), it is clear that $(x, u(x_1, \ldots, x_n)) \in \theta = \ker(h)$ if and only if

$$u(x, x, \ldots, x) = h(u(x_1, \ldots, x_n)) = h(x) = x$$

in $\mathbf{F}$, which holds if and only if u is idempotent. Thus $B = x/\theta$, as claimed. If $u(x_1, \ldots, x_n)$ and $v(x_1, \ldots, x_n)$ are arbitrarily chosen idempotent terms, then using the term

$$u(v(x_{11}, \ldots, x_{1n}), \ldots, v(x_{n1}, \ldots, x_{nn}))\,,$$

and the fact that $(x_i, x_j) \in E \subseteq T$ for all i and j, we derive that

$$\begin{aligned} u(x_1, \ldots, x_n) &= u(v(x_1, \ldots, x_1), \ldots, v(x_n, \ldots, x_n)) \\ &\equiv_T u(v(x_1, \ldots, x_n), \ldots, v(x_1, \ldots, x_n)) \\ &= v(x_1, \ldots, x_n)\,. \end{aligned}$$

Thus B is contained in a T-block. That it is the unique T-block containing x follows from the fact that $B = x/\theta$ is a θ-block (so no element in $F - B$ is T-related to any element of B). □

LEMMA 3.7. *Let $\mathcal{V}$ be a variety and $\mathbf{F} = \mathbf{F}_{\mathcal{V}}(x, y)$ the free $\mathcal{V}$-algebra generated by $\{x, y\}$. Let $\theta = \mathrm{Cg}^{\mathbf{F}}(x, y)$ and T the tolerance of $\mathbf{F}$ generated by (x, y). Let $B = x/\theta$.*

(1) *Let ρ be a congruence on $\mathbf{F}$ such that $\mathbf{SR}(T, T; \rho)$ holds. If*

$$\begin{bmatrix} p & q \\ r & s \end{bmatrix} \in M(\theta, \theta)$$

and $p, q, r, s \in B$, then $q \equiv_\rho r \implies r \equiv_\rho s$.

(2) *Let ρ be a congruence on $\mathbf{F}$ such that $\mathbf{S}(T, T; \rho)$ holds. If*

$$\begin{bmatrix} p & q \\ r & s \end{bmatrix} \in M(\theta, \theta)$$

and $p, q, r, s \in B$, then $p \equiv_\rho q \implies r \equiv_\rho s$.

PROOF. Choose any θ, θ-matrix

$$\begin{bmatrix} p & q \\ r & s \end{bmatrix} = \begin{bmatrix} t(\mathbf{a}, \mathbf{c}) & t(\mathbf{a}, \mathbf{d}) \\ t(\mathbf{b}, \mathbf{c}) & t(\mathbf{b}, \mathbf{d}) \end{bmatrix}.$$

Let $h\colon \mathbf{F} \to \mathbf{F}$ be the endomorphism determined by $x \mapsto x$, $y \mapsto x$. The assumption $\mathbf{a} \equiv_\theta \mathbf{b}$ is equivalent to $h(\mathbf{a}) = h(\mathbf{b})$, and similarly $\mathbf{c} \equiv_\theta \mathbf{d}$ is equivalent to

$h(\mathbf{c}) = h(\mathbf{d})$ (where by $h(\mathbf{a})$, etc., we mean that we apply the function h componentwise). Consider the following 4×4 matrix:

$$\begin{bmatrix} t(\mathbf{a},\mathbf{c}) & t(\mathbf{a},h(\mathbf{c})) & = & t(\mathbf{a},h(\mathbf{d})) & t(\mathbf{a},\mathbf{d}) \\ t(h(\mathbf{a}),\mathbf{c}) & t(h(\mathbf{a}),h(\mathbf{c})) & = & t(h(\mathbf{a}),h(\mathbf{d})) & t(h(\mathbf{a}),\mathbf{d}) \\ \| & \| & & \| & \| \\ t(h(\mathbf{b}),\mathbf{c}) & t(h(\mathbf{b}),h(\mathbf{c})) & = & t(h(\mathbf{b}),h(\mathbf{d})) & t(h(\mathbf{b}),\mathbf{d}) \\ t(\mathbf{b},\mathbf{c}) & t(\mathbf{b},h(\mathbf{c})) & = & t(\mathbf{b},h(\mathbf{d})) & t(\mathbf{b},\mathbf{d}) \end{bmatrix}.$$

We claim that the four elements in the middle are equal to x. Indeed, the assumption $p, q, r, s \in B$ implies $h(p) = h(q) = h(r) = h(s) = x$. Therefore, for example,

$$t(h(\mathbf{b}), h(\mathbf{d})) = h(t(\mathbf{b}, \mathbf{d})) = h(s) = x\,.$$

Note also that the four 2×2 matrices in the corners are T,T-matrices, since $\mathbf{u}$ and $h(\mathbf{u})$ are T-related componentwise by Lemma 3.6 (2). All elements of this 4×4 matrix are contained in B, because all elements are θ-related, the corners are in B by assumption, and B is a block of θ.

Next, let $k\colon \mathbf{F} \to \mathbf{F}$ be the endomorphism of $\mathbf{F}$ that sends x to y and y to y. Consider the analogous 4×4 matrix with respect to k. All the statements above remain true with k in place of h, with the exception that in this case the four elements in the middle are equal to y, not x. Now apply $q(x, y)$ to these two matrices, the first matrix in the first argument, the second matrix in the second argument. This produces a 4×4 matrix which, after deleting one of the doubled middle columns and one of the doubled middle rows, is a 3×3 matrix with certain known entries:

$$J = \begin{bmatrix} p & i & q \\ j & q & \ell \\ r & m & s \end{bmatrix}.$$

Indeed, the middle element is $q(x, y) = q$, the element in the upper left corner is $q(p, p) = p$, since q is idempotent, and a similar argument proves that in the other corners we get q, r and s. It will not matter in this argument what i, j, ℓ, m are. The four 2×2 matrices in the corners of the 3×3 matrix J are still T,T-matrices.

To prove (1), we have assumed that $q \equiv_\rho r$. We have also assumed that $\mathbf{SR}(T, T; \rho)$ holds, so from the T,T-matrix $\begin{bmatrix} i & q \\ q & \ell \end{bmatrix}$ in the upper right corner of J we derive that $q \equiv_\rho \ell$. From the matrix $\begin{bmatrix} j & q \\ r & m \end{bmatrix}$ in the lower left corner of J (and the assumption that $q \equiv_\rho r$) we get that $r \equiv_\rho m$. Thus $\ell \equiv_\rho q \equiv_\rho r \equiv_\rho m$, so since the cross diagonal entries of the T,T-matrix $\begin{bmatrix} q & \ell \\ m & s \end{bmatrix}$ in the lower right corner are ρ-related we deduce that $m \equiv_\rho s$. Thus $r \equiv_\rho m \equiv_\rho s$, and (1) is proved.

To prove (2), we have assumed that $p \equiv_\rho q$. We have also assumed that $\mathbf{S}(T, T; \rho)$ holds, and therefore $\mathbf{SR}(T, T; \rho)$ and $\mathbf{C}(T, T; \rho)$ both hold. From the T,T-matrix $\begin{bmatrix} i & q \\ q & \ell \end{bmatrix}$ in the upper right corner of J we get by strong rectangulation that $q \equiv_\rho \ell$. From the T,T-matrix $\begin{bmatrix} i & p \\ q & j \end{bmatrix}$, which is obtained from the T,T-matrix in the upper left corner of J by switching columns, we get (from $p \equiv_\rho q$)

that $q \equiv_\rho j$. Now that we have $j \equiv_\rho q$ and $q \equiv_\rho \ell$, we apply $\mathbf{C}(T, T; \rho)$ to the matrices

$$\begin{bmatrix} j & q \\ r & m \end{bmatrix} \quad \text{and} \quad \begin{bmatrix} q & \ell \\ m & s \end{bmatrix}$$

from the bottom two corners of J to obtain $r \equiv_\rho m$ and $m \equiv_\rho s$. Therefore $r \equiv_\rho m \equiv_\rho s$, and (2) is proved. □

Our next task is to describe how to construct the $\mathbf{Q}$-commutator $[S, T]_\mathbf{Q}$ of two tolerances where $\mathbf{Q} = \mathbf{S}$ or $\mathbf{SR}$. Fix the choice of $\mathbf{Q} \in \{\mathbf{S}, \mathbf{SR}\}$. Let $\mathbf{A}$ be an algebra, and let S and T be tolerances of $\mathbf{A}$. Inductively define binary relations τ_n on $\mathbf{A}$ as follows. Let τ_0 be the equality relation. If τ_n is defined, let τ_{n+1} be the symmetric and transitive closure of the set of all pairs that can be obtained in one of the following two ways. For each matrix

$$\begin{bmatrix} p & q \\ r & s \end{bmatrix} \in M(S, T),$$

(1) if $\mathbf{Q} = \mathbf{S}$ or $\mathbf{SR}$ and $q \equiv_{\tau_n} r$, then put the pair (r, s) into τ_{n+1};
(2) if $\mathbf{Q} = \mathbf{S}$ and $p \equiv_{\tau_n} q$, then also put the pair (r, s) into τ_{n+1}.

Let $\tau = \bigcup_{n<\omega} \tau_n$.

LEMMA 3.8. $[S, T]_\mathbf{Q} = \tau$.

PROOF. The set of S, T-matrices is closed under the componentwise application of unary polynomials of $\mathbf{A}$, so inductively the same property holds for every τ_n. Since τ_n is a symmetric, transitive relation and is closed under the application of unary polynomials, it is a congruence.

CLAIM 3.9. *$\tau_n \leq \tau_{n+1}$ for all n.*

The argument is by induction. It is clear that $\tau_0 \leq \tau_1$, since τ_0 is the equality relation. Suppose that $\tau_k \leq \tau_{k+1}$ holds for some k. Suppose that (r, s) is a generator of τ_{k+1} because, say, there is some

$$\begin{bmatrix} p & q \\ r & s \end{bmatrix} \in M(S, T),$$

such that $q \equiv_{\tau_k} r$ (reason (1) preceding the statement of the lemma). Then since $\tau_k \leq \tau_{k+1}$ we have $q \equiv_{\tau_{k+1}} r$, which shows that the same matrix witnesses that (r, s) is a generator for τ_{k+2}. Since τ_{k+2} contains the generators of τ_{k+1}, we have $\tau_{k+1} \leq \tau_{k+2}$, proving the claim.

Since $\tau_n \leq \tau_{n+1}$ for every n, the union $\tau = \bigcup_{n<\omega} \tau_n$ is also a congruence on $\mathbf{A}$.

Let $\delta = [S, T]_\mathbf{Q}$. We argue by induction that $\tau_n \leq \delta$ for all n. This is trivially true if $n = 0$, so suppose that $\tau_n \leq \delta$ for some fixed n and that (r, s) is one of the generators of τ_{n+1}. If $(r, s) \in \tau_{n+1}$ because of reason (1) preceding the statement of this lemma, then there is a matrix

$$\begin{bmatrix} p & q \\ r & s \end{bmatrix} \in M(S, T),$$

such that $q \equiv_{\tau_n} r$. We have $q \equiv_\delta r$ (since $\tau_n \leq \delta$) and also $\mathbf{Q}(S, T; \delta)$ (hence $\mathbf{SR}(S, T; \delta)$) since $\delta = [S, T]_\mathbf{Q}$. Therefore we have $(r, s) \in \delta$. If $(r, s) \in \tau_{n+1}$ because of reason (2), then we must have $\mathbf{Q} = \mathbf{S}$ and so $\mathbf{C}(S, T; \delta)$ holds. Now a similar argument shows that $(r, s) \in \delta$. Thus δ contains all the generators of τ_{n+1}, so $\tau_{n+1} \leq \delta$. By induction $\tau_n \leq \delta$ for all n, so $\tau \leq \delta = [S, T]_\mathbf{Q}$.

To finish the proof we must verify that $[S,T]_{\mathbf{Q}} \leq \tau$, and for this it suffices to show that $\mathbf{Q}(S,T;\tau)$ holds. Choose any S,T-matrix

$$\begin{bmatrix} p & q \\ r & s \end{bmatrix},$$

and assume that $q \equiv_\tau r$. Then $q \equiv_{\tau_n} r$ for some n, so from item (1) preceding the lemma we have $(r,s) \in \tau_{n+1} \leq \tau$. Thus $\mathbf{SR}(S,T;\tau)$ holds, and $[S,T]_{\mathbf{Q}} = \tau$ when $\mathbf{Q} = \mathbf{SR}$. If $\mathbf{Q} = \mathbf{S}$, then a similar argument using (2) in place of (1) shows that $\mathbf{C}(S,T;\tau)$ holds, so $\mathbf{S}(S,T;\tau)$ holds (and so $[S,T]_{\mathbf{Q}} = \tau$ when $\mathbf{Q} = \mathbf{S}$). □

Now we can prove that the four conditions of Lemma 3.5 are equivalent.

LEMMA 3.10. *Let $\mathcal{V}$ be any variety, and let $\mathbf{Q} = \mathbf{S}$ or $\mathbf{SR}$. The following are equivalent.*

(1) *Some algebra in $\mathcal{V}$ has a nonzero $\mathbf{Q}$-abelian congruence.*
(2) *Some algebra in $\mathcal{V}$ has a nonzero $\mathbf{Q}$-abelian tolerance.*

PROOF. Denote by $\mathbf{F}$ the free $\mathcal{V}$-algebra generated by $\{x,y\}$, by T the tolerance of $\mathbf{F}$ generated by (x,y), and by θ the congruence of $\mathbf{F}$ generated by (x,y). Let $B = x/\theta$. According to Lemma 3.5, we must establish that $(x,y) \notin [\theta,\theta]_{\mathbf{Q}}$ if and only if $(x,y) \notin [T,T]_{\mathbf{Q}}$. Since $x,y \in B$, this will follow if we show that the restrictions of the relevant congruences to B are equal, i.e., if we show that,

$$[\theta,\theta]_{\mathbf{Q}}|_B = [T,T]_{\mathbf{Q}}|_B\,.$$

We proceed to do this.

Since $T \subseteq \theta$ we have $[T,T]_{\mathbf{Q}}|_B \subseteq [\theta,\theta]_{\mathbf{Q}}|_B$, so we only need to verify the reverse inclusion. Let $\rho = [T,T]_{\mathbf{Q}}$. Build $\tau = [\theta,\theta]_{\mathbf{Q}}$ according to the recipe described preceding Lemma 3.8. We argue by induction that $\tau_n|_B \subseteq \rho$. This is clear for $n = 0$. Suppose that $n \geq 0$ and that $\tau_n|_B \subseteq \rho$. Choose any $(u,v) \in \tau_{n+1}|_B$. By the definition of τ_{n+1}, there exists a chain $u = u_1, u_2, \ldots, u_k = v$ such that for every $1 \leq i < k$ the pair (u_i, u_{i+1}) or the pair (u_{i+1}, u_i) is produced by rule (1) or (2) of the definition of τ_{n+1}. That is, there exists a θ,θ-matrix

$$\begin{bmatrix} p & q \\ r & s \end{bmatrix}$$

such that $\{u_i, u_{i+1}\} = \{r,s\}$, and either (q,r) or (p,q) is contained in τ_n. We want to show that $u_i \equiv_\rho u_{i+1}$.

As $u \in B$ and $\tau \leq \theta$, every u_i belongs to B. Hence the entries of the θ,θ-matrix above are in B, too. In the first case, when $(q,r) \in \tau_n$ we therefore have that $(q,r) \in \rho$ by the induction assumption. Then (1) of Lemma 3.7 shows that $(r,s) \in \rho$ as desired. In the second case, when $(p,q) \in \tau_n$ (which only arises when $\mathbf{Q} = \mathbf{S}$) we use (2) of Lemma 3.7 to reach the same conclusion. Thus the induction step, and hence the proof of the lemma is complete. □

The next task we set for ourselves is to show that a variety has no member with a strongly abelian tolerance if and only if the variety has a Taylor term. To prove this we need a description of the T,T-matrices where T is the tolerance on $\mathbf{F}_{\mathcal{V}}(x,y)$ generated by (x,y).

LEMMA 3.11. *Let $\mathcal{V}$ be a variety, $\mathbf{F} = \mathbf{F}_\mathcal{V}(x, y)$ the free $\mathcal{V}$-algebra generated by $\{x, y\}$, and let T be the tolerance of $\mathbf{F}$ generated by (x, y). The T, T-matrices in $\mathbf{F}$ are exactly the matrices of the form*

$$\begin{bmatrix} f(x,y,x,y,x,y) & f(x,x,x,y,y,y) \\ f(y,y,x,x,x,y) & f(y,x,x,x,y,y) \end{bmatrix}$$

where f is a sixary term.

PROOF. A typical T, T-matrix has the form

$$\begin{bmatrix} g(\mathbf{a},\mathbf{c}) & g(\mathbf{a},\mathbf{d}) \\ g(\mathbf{b},\mathbf{c}) & g(\mathbf{b},\mathbf{d}) \end{bmatrix}$$

where $(a_i, b_i) \in T$, $(c_j, d_j) \in T$, and (we may assume) g is a term. Since T is the subuniverse of F^2 generated by (x, y), (y, x), (x, x) and (y, y), a typical pair in T has the form

$$r\big((x,y),(y,x),(x,x),(y,y)\big) = \big(r(x,y,x,y), r(y,x,x,y)\big)$$

for some 4-ary term r. Therefore a typical T, T-matrix may be written as

$$\begin{bmatrix} g\big(\mathbf{r}(x,y,x,y),\mathbf{s}(x,y,x,y)\big) & g\big(\mathbf{r}(x,y,x,y),\mathbf{s}(y,x,x,y)\big) \\ g\big(\mathbf{r}(y,x,x,y),\mathbf{s}(x,y,x,y)\big) & g\big(\mathbf{r}(y,x,x,y),\mathbf{s}(y,x,x,y)\big) \end{bmatrix}.$$

If we define $f(x_1,\ldots,x_6) := g\big(\mathbf{r}(x_1,x_4,x_3,x_6),\mathbf{s}(x_5,x_2,x_3,x_6)\big)$, then this matrix is the one described in the lemma. □

LEMMA 3.12. *If no algebra in the variety $\mathcal{V}$ has a nonzero strongly abelian tolerance, then $\mathcal{V}$ satisfies a nontrivial idempotent Maltsev condition.*

PROOF. Let $\mathbf{F}$ be the free $\mathcal{V}$-algebra generated by $\{w, z\}$, let T be the tolerance of $\mathbf{F}$ generated by (w, z), and let B be the T-block $w/\mathrm{Cg}^{\mathbf{F}}(w, z)$. For each element $u \in F$ choose and fix a binary term $u(x, y)$ such that $u(w, z) = u$. Build $\tau = \bigcup_{n<\omega} \tau_n = [T, T]_\mathbf{S}$ using the method described before Lemma 3.8. For each pair $(u, v) \in \tau$ we shall define a finite set of equations $\Sigma(u, v)$. If $(u, v) \in \tau_0$, let $\Sigma(u, v) = \{u(x, y) \approx v(x, y)\}$. To define $\Sigma(u, v)$ when $(u, v) \in \tau_{n+1} - \tau_n$, choose and fix a finite sequence $u = u_1, u_2, \ldots, u_{k+1} = v$ of elements of F such that for each i either (u_i, u_{i+1}) or (u_{i+1}, u_i) is a generating pair for τ_{n+1}. Generating pairs for τ_{n+1} come from T, T-matrices, and we know from Lemma 3.11 that the T, T-matrices in $\mathbf{F}$ are exactly the matrices of the form

$$\begin{bmatrix} p & q \\ r & s \end{bmatrix} = \begin{bmatrix} f(w,z,w,z,w,z) & f(w,w,w,z,z,z) \\ f(z,z,w,w,w,z) & f(z,w,w,w,z,z) \end{bmatrix}.$$

Therefore for each i between 1 and k there is a sixary term f_i such that

$$(3.2)\qquad \{u_i, u_{i+1}\} = \{r_i, s_i\} = \{f_i(z,z,w,w,w,z), f_i(z,w,w,w,z,z)\}$$

where one of the following "side conditions" holds:

$$(3.3)\qquad \big(f_i(w,w,w,z,z,z), f_i(z,z,w,w,w,z)\big) = (q_i, r_i) \in \tau_n, \quad \text{or}$$

$$(3.4)\qquad \big(f_i(w,z,w,z,w,z), f_i(w,w,w,z,z,z)\big) = (p_i, q_i) \in \tau_n.$$

Let $(r_i^\circ, s_i^\circ) := (r_i, s_i)$ if $(r_i, s_i) = (u_i, u_{i+1})$ and let $(r_i^\circ, s_i^\circ) := (s_i, r_i)$ if $(r_i, s_i) = (u_{i+1}, u_i)$. In either case, $r_i^\circ = u_i$ and $s_i^\circ = u_{i+1}$ in $\mathbf{F}$.

Define $\Sigma(u, v)$ to be the set of the following identities.

(i) $u(x, y) \approx r_1^\circ(x, y)$ and $s_k^\circ(x, y) \approx v(x, y)$.

(ii) $s_i^\circ(x, y) \approx r_{i+1}^\circ(x, y)$ for each $1 \le i \le k$.

(iii) $p_i(x,y) \approx f_i(x,y,x,y,x,y)$, $q_i(x,y) \approx f_i(x,x,x,y,y,y)$,
$r_i(x,y) \approx f_i(y,y,x,x,x,y)$, and $s_i(x,y) \approx f_i(y,x,x,x,y,y)$.

(iv) The identities in $\Sigma(q_i, r_i)$ for each i whose side condition is (3.3), along with
$q_i(x,y) \approx f_i(x,x,x,y,y,y)$ and
$r_i(x,y) \approx f_i(y,y,x,x,x,y)$.

(v) The identities in $\Sigma(p_i, q_i)$ for each i whose side condition is (3.4), along with
$p_i(x,y) \approx f_i(x,y,x,y,x,y)$ and
$q_i(x,y) \approx f_i(x,x,x,y,y,y)$.

The identities in (i) and (ii) express that the pairs $\big(r_i^\circ(w,z), s_i^\circ(w,z)\big)$ form a directed chain connecting u to v, the identities in (iii) express that $\big(r_i^\circ(w,z), s_i^\circ(w,z)\big)$ is the bottom row of a T,T-matrix, while the identities of type (iv) and (v) express that the relevant T,T-matrices satisfy side conditions sufficient to guarantee that $\big(r_i(w,z), s_i(w,z)\big)$ is a generator of τ_{n+1}. These identities hold in $\mathcal{V}$, since they become equalities if we substitute the free generators w and z of $\mathbf{F}$ for the variables x and y.

Induction on n may be used to prove that $\Sigma(u,v)$ is a finite set of identities for every $(u,v) \in \tau_{n+1} - \tau_n$, and that if $u \in B$ then all terms appearing in the identities in $\Sigma(u,v)$ are idempotent. Moreover, induction on n also shows that if $(u,v) \in \tau$, then for any pair (a,b) from any tolerance S on an algebra $\mathbf{A} \in \mathcal{V}$ the pair $\big(u(a,b), v(a,b)\big)$ belongs to the strong commutator $[S,S]_{\mathbf{S}}$.

We have assumed that no algebra in $\mathcal{V}$ has a nontrivial strongly abelian tolerance, so by Lemma 3.5 we have $(w,z) \in [T,T]_{\mathbf{S}} = \bigcup_{n<\omega} \tau_n$, that is, $(w,z) \in \tau_{n+1} - \tau_n$ for some n. Therefore $\mathcal{V}$ satisfies the finite set of idempotent identities in $\Sigma(w,z)$. These identities constitute an idempotent Maltsev condition which, when satisfied by a variety $\mathcal{W}$, forces the following property: for any pair (c,d) from any tolerance R on an algebra $\mathbf{B} \in \mathcal{W}$ the pair $\big(w(c,d), z(c,d)\big) = (c,d)$ belongs to the strong commutator $[R,R]_{\mathbf{S}}$. Using the monotonicity of the strong commutator (Theorem 3.4 (1)), this is equivalent to the property that $(c,d) \in [\mathrm{Tg}^{\mathbf{B}}(c,d), \mathrm{Tg}^{\mathbf{B}}(c,d)]_{\mathbf{S}}$. Using Lemma 3.5 again we derive that if this Maltsev condition holds in $\mathcal{W}$, then no algebra in $\mathcal{W}$ has a nontrivial strongly abelian tolerance. Since every tolerance on every member of the variety of sets is strongly abelian, this idempotent Maltsev condition fails in the variety of sets, hence is nontrivial. □

THEOREM 3.13. *Let $\mathcal{V}$ be a variety. The following are equivalent.*

(1) *$\mathcal{V}$ has no member with a nonzero strongly abelian congruence.*
(2) *$\mathcal{V}$ has no member with a nonzero strongly abelian tolerance.*
(3) *$\mathcal{V}$ satisfies a nontrivial idempotent Maltsev condition.*

PROOF. We proved in Lemma 3.10 that items (1) and (2) are equivalent, and in Lemma 3.12 that (2) $\Longrightarrow$ (3). It remains to prove that (3) $\Longrightarrow$ (1).

Assume that (3) holds, and let $f(x_1,\dots,x_n)$ be a Taylor term for $\mathcal{V}$. Arbitrarily choose and fix an algebra $\mathbf{A} \in \mathcal{V}$, a strongly abelian congruence $\alpha \in \mathrm{Con}(\mathbf{A})$, and a pair $(a,b) \in \alpha$. Let $f_i(x,\mathbf{y})$ be an n-ary term obtained from f by reordering the variables so that x_i is first. (That is, $f(x_1,\dots,x_n) = f_i(x_i,\mathbf{y})$ where $\mathbf{y}$ is an ordering of $\{x_1,\dots,x_n\} \setminus \{x_i\}$.) The i-th Taylor identity implies that $f_i(a,\mathbf{u}) = f_i(b,\mathbf{v})$ for some $(n-1)$-tuples $\mathbf{u}$ and $\mathbf{v}$ consisting of a's and b's. Therefore

$$\begin{bmatrix} p & q \\ r & s \end{bmatrix} = \begin{bmatrix} f_i(a,\mathbf{v}) & f_i(b,\mathbf{v}) \\ f_i(a,\mathbf{u}) & f_i(b,\mathbf{u}) \end{bmatrix} \in M(\alpha,\alpha)$$

is an α,α-matrix with $q=r$. Since α is strongly abelian it is strongly rectangular, so $f_i(a,\mathbf{u}) = r = s = f_i(b,\mathbf{u})$. Now choose $\mathbf{w} \in \{a,b\}^{n-1}$ arbitrarily. The matrix

$$\begin{bmatrix} P & Q \\ R & S \end{bmatrix} = \begin{bmatrix} f_i(a,\mathbf{u}) & f_i(b,\mathbf{u}) \\ f_i(a,\mathbf{w}) & f_i(b,\mathbf{w}) \end{bmatrix} \in M(\alpha,\alpha)$$

is an α,α-matrix with $P=Q$. Since α is strongly abelian, it is abelian, so $f_i(a,\mathbf{w}) = R = S = f_i(b,\mathbf{w})$.

What this says about the original Taylor term is that, for every $1 \le i \le n$, if $\mathbf{c},\mathbf{d} \in \{a,b\}^n$ are tuples that agree in every coordinate except possibly the i-th, then $f(\mathbf{c}) = f(\mathbf{d})$. This statement is true for all i, so

$$\begin{aligned} a &= f(a,a,a,\dots,a) \\ &= f(b,a,a,\dots,a) \\ &= f(b,b,a,\dots,a) \\ &\;\;\vdots \\ &= f(b,b,b,\dots,b) = b, \end{aligned}$$

forcing $a=b$. Since (a,b) was chosen arbitrarily, we conclude that no $\mathbf{A} \in \mathcal{V}$ has a nontrivial strongly abelian congruence. □

3.2. Join Terms

In the previous section we showed that a variety omits strongly abelian congruences or tolerances if and only if it satisfies a nontrivial idempotent Maltsev condition (Theorem 3.13). In this section we will prove that the varieties with these properties are exactly the varieties that omit strongly rectangular congruences or tolerances. The obvious approach to this result is to note that if a variety omits strongly rectangular congruences or tolerances, then it must omit strongly abelian congruences or tolerances, and therefore must satisfy a nontrivial idempotent Maltsev condition. Thus, the only thing to show is that it is possible to modify the proof of Theorem 3.13 (3) $\Longrightarrow$ (1) so that it only uses the assumption of strong rectangulation.

It must be possible to obtain the result in this way, but in this section we choose another path. We instead develop machinery for dealing with stongly rectangular congruences and tolerances that makes the proof of the desired result a triviality. Although the path we choose does not lead to the result more quickly or more simply than the path sketched in the previous paragraph, we feel that developing the machinery for strong rectangulation is more important than the target result.

Definition 3.14. Let $\mathbf{A}$ be an algebra, let T be a tolerance on $\mathbf{A}$ and let δ be a congruence on $\mathbf{A}$. A (T,T)**-triple** is a triple $(b,c;d)$ such that there is a matrix

$$\begin{bmatrix} a & b \\ c & d \end{bmatrix} \in M(T,T)\,.$$

A $(T,T;\delta)$**-pair** is a pair $(c,d) \in A^2$ such that there is a (T,T)-triple $(b,c;d)$ with $b \equiv_\delta c$.

Observe that the set of all (T,T)-triples is a subuniverse of $\mathbf{A}^3$, and the set of all $(T,T;\delta)$-pairs is a subuniverse of $\mathbf{A}^2$.

DEFINITION 3.15. Let T be a tolerance on an algebra $\mathbf{A}$. For any block B of T write $a \geq_B b$ (or $b \leq_B a$) to mean that there exist $(T,T;0)$-pairs $(u_1,u_2),\ldots,(u_{n-1},u_n)$ such that $a = u_1$, $b = u_n$, and $u_i \in B$ for all i. Write $a \sim_B b$ to mean $a \geq_B b$ and $a \leq_B b$.

Write $a \geq_T b$ (or $b \leq_T a$) if there is some T-block B for which $a \geq_B b$. Similarly, write $a \sim_T b$ if $\exists B(a \sim_B b)$.

There is a potential ambiguity in the notation $\geq_B$, because a subset B may be a block of more than one tolerance, but this issue will not arise in this monograph.

LEMMA 3.16. *Let T be a tolerance on an algebra $\mathbf{A}$. The following hold.*

(1) *If B is a block of T, then $\geq_B$ is a quasi-order on B, and $\sim_B$ is the induced equivalence relation on B. Both relations are compatible with the idempotent polynomials of $\mathbf{A}$.*
(2) *The relation $\geq_T$ is a reflexive compatible relation of $\mathbf{A}$, and $\sim_T$ is a tolerance on $\mathbf{A}$.*
(3) *If $\mathbf{SR}(T,T;0)$ holds, then $\geq_T$ is the equality relation on $\mathbf{A}$.*
(4) *If B is a block of T and is also a block of $\mathrm{Cg}^{\mathbf{A}}(T)$, then $\geq_B$ equals the restriction to B of $\geq_T$ and also equals the restriction to B of the transitive closure of the set of $(T,T;0)$-pairs.*
(5) *If T is a congruence, then $\geq_T$ is a quasi-order of $\mathbf{A}$, which is the transitive closure of the set of $(T,T;0)$-pairs. The relation $\sim_T$ is the equivalence relation induced by $\geq_T$, which is a congruence on $\mathbf{A}$ contained in T.*

PROOF. For this proof, let S denote the set of $(T,T;0)$-pairs of $\mathbf{A}$.

For item (1), the matrix $\begin{bmatrix} u & u \\ u & u \end{bmatrix}$ belongs to $M(T,T)$ for any $u \in A$, so the trivial pair (u,u) is a $(T,T;0)$-pair for any $u \in A$. This is enough to show that $\geq_B$ is reflexive on B. It is clear from Definition 3.15 that $\geq_B$ equals tr.cl.$(S|_B)$, hence $\geq_B$ is transitive as well. This proves that $\geq_B$ is a quasi-order on B; it follows from Definition 3.15 that $\sim_B$ is the induced equivalence relation. If $p(x_1,\ldots,x_n)$ is an idempotent polynomial of $\mathbf{A}$, then $p(B,\ldots,B)$ is a subset of A containing B and consisting of pairwise T-related elements. Since B is a block, $p(B,\ldots,B) = B$. As S is compatible with all polynomials of $\mathbf{A}$, the fact that $p(B,\ldots,B) = B$ implies that $S|_B$ is compatible with p. Therefore $\geq_B=$ tr.cl.$(S|_B)$ is also compatible with p, as is the intersection $\sim_B$ of this relation with its converse.

Now we prove item (2). Since $\geq_T$ equals the union over all blocks B of the relations $\geq_B$, it follows that $\geq_T$ is reflexive. Similarly, $\sim_T$ is reflexive and symmetric. It is only necessary to prove that $\geq_T$ and $\sim_T$ are compatible relations. We will prove the following slightly stronger fact.

CLAIM 3.17. *Let $f(x_1,\ldots,x_n)$ be an n-ary basic operation of $\mathbf{A}$. Assume that $B_1,\ldots,B_n,B$ are T-blocks such that $f(B_1,\ldots,B_n) \subseteq B$. If $a_i \geq_{B_i} b_i$ holds for all i, then*

$$f(a_1,\ldots,a_n) \geq_B f(b_1,\ldots,b_n)\,.$$

The same statement is true with $\sim$ in place of $\geq$.

The arguments are essentially the same for $\geq$ and $\sim$, so we prove the claim only for $\geq$. By our assumption $a_i \geq_{B_i} b_i$, so there exist $(T,T;0)$-pairs $(u_i^1,u_i^2),\ldots,$

This shows that $a \cup b$ is $\widehat{T}$-related to any $c \in D$, so $a \cup b \in D$ by the fact that D is a block. Since $\cup$ is a homomorphism from $\widehat{T}$ to $\widehat{\mathbf{A}}$ that is the restriction of a semilattice operation on $\mathbf{B}/\theta$, and since each $\widehat{T}$-block is closed under $\cup$, it follows that $\cup$ is a compatible semilattice operation on $\widehat{T}$.

Now we prove that $\widehat{T}$ is rectangular. To check this we need a compatible partial order on $\widehat{\mathbf{A}}$. We choose it to be the restriction of the $\cup$-semilattice order on $\mathbf{B}/\theta$, which we denote by $\gg$ and define by

$$a \gg b \iff a = a \cup b\,.$$

We have already observed that the operations of $\mathbf{B}/\theta$ are monotone with respect to the semilattice order, and this fact is inherited by $\widehat{\mathbf{A}}$, hence $\gg$ is a compatible partial order. To see that $\mathbf{R}(\widehat{T},\widehat{T};\gg)$ holds, choose $M \in M(\widehat{T},\widehat{T})$,

$$M = \begin{bmatrix} p(\mathbf{a},\mathbf{u}) & p(\mathbf{a},\mathbf{v}) \\ p(\mathbf{b},\mathbf{u}) & p(\mathbf{b},\mathbf{v}) \end{bmatrix} = \begin{bmatrix} p & q \\ r & s \end{bmatrix},$$

with $u \gg q, r$ for some $u \in \widehat{A}$. All four elements of this matrix are $\widehat{T}$-related, so

$$u \gg u \cup u \gg q \cup r = p(\mathbf{a},\mathbf{v}) \cup p(\mathbf{b},\mathbf{u}) = p(\mathbf{a}\cup\mathbf{b},\mathbf{v}\cup\mathbf{u}) \gg p(\mathbf{b},\mathbf{v}) = s\,.$$

Thus, $u \gg q, r$ implies $u \gg s$, as required. This completes the proof of the claim, and also of the theorem. □

It is worth recording that the proof produces from $\mathbf{A}$ and T two algebras of interest: $\widehat{\mathbf{A}}$ and $\mathbf{B}/\theta$. The properties of the two algebras are roughly the same with respect to rectangulation, but $\widehat{\mathbf{A}}$ is in the variety generated by $\mathbf{A}$ while $\mathbf{B}/\theta$ may not be. On the other hand, the compatible semilattice operation that we constructed is only a partial operation on $\widehat{\mathbf{A}}$ (it is defined only on $\widehat{T}$-blocks), while it is a total operation on $\mathbf{B}/\theta$ and the semilattice order is a compatible partial order of $\mathbf{B}/\theta$.

5.2. Rectangular Tolerances and Join Terms

We introduced join terms in Section 3.2, but did not investigate their properties. It is now important that we do so. In this section we explore the structure of rectangular tolerances on algebras in varieties with a join term. For such tolerances we prove that the representation given in Theorem 5.7 is simpler in that one may take $\widehat{\mathbf{A}} = \mathbf{A}$ and $\widehat{T} = T$. We then describe connections with the relations $\geq_T$ and $\sim_T$ that were introduced in Definition 3.15

THEOREM 5.13. *Let $\mathcal{V}$ be a variety with a join term $+$, and let T be a tolerance on an algebra $\mathbf{A} \in \mathcal{V}$.*

(1) *If B is a T-block and $w \geq_B u$ and $w \geq_B v$, then $w \geq_B u + v$. Similarly, if $w \geq_T u$ and $w \geq_T v$, then $w \geq_T u + v$.*
(2) *If B is a T-block, then $\geq_B$ is a join semilattice ordering on B modulo $\sim_B$, and $+$ is the join with respect to that ordering.*
(3) *If B is a T-block containing all entries of $\begin{bmatrix} a & b \\ c & d \end{bmatrix} \in M(T,T)$, then $a + d \sim_B b + c$.*
(4) *If $(q, r; s)$ is a (T,T)-triple, then $q + r \geq_T s$.*
(5) *If p is a polynomial of $\mathbf{A}$ and $\mathbf{u}\, T\, \mathbf{v}$, then*

$$p(\mathbf{u}+\mathbf{v}) \sim_T p(\mathbf{u}) + p(\mathbf{v})\,.$$

PROOF. From $w \geq_B u$ and $w \geq_B v$ we get $w = w + w \geq_B u + v$, since $+$ is idempotent and $\geq_B$ is compatible with idempotent polynomials (Lemma 3.16 (1)). The statement about $\geq_T$ follows in the same way from Lemma 3.16 (2).

If B is a T-block, then B is closed under $+$ since $+$ is idempotent. From Corollary 3.22, $u + v \geq_B u, v$, while item (1) proves conversely that if $w \geq_B u, v$, then $w \geq_B u + v$. Thus $\geq_B$ is indeed a semilattice order modulo $\sim_B$ on B, and $+$ is the join with respect to this order.

For item (3), Lemma 3.18 (3) proves that $b + c \geq_B (b + d) + (a + c)$. But from Corollary 3.22 we have $b + d \geq_B d$ and $a + c \geq_B a$, so

$$b + c \geq_B (b + d) + (a + c) \geq_B d + a\,.$$

Since $\geq_B$ is transitive on B, we have $b + c \geq_B d + a$. If we interchange the two rows of this T, T-matrix, then we derive the reverse inequality $d + a \geq_B b + c$, hence $b + c \sim_B d + a$. It follows from item (2) that $d + a \sim_B a + d$, so from the transitivity of $\sim_B$ on B we get $a + d \sim_B b + c$, as claimed.

For (4), if $(q, r; s)$ is a (T, T)-triple, then its entries belong to a T, T-matrix. All four entries of any T, T-matrix are pairwise T-related, hence a block B exists that contains all four entries. Applying item (3) to this block we obtain $q + r \sim_B p + s$ for some $p \in B$, hence from item (1) it follows that $q + r \geq_B s$. This proves $q + r \geq_T s$.

We first prove item (5) in the case where $p = f$ is a basic operation of $\mathbf{A}$. Choose T-blocks B_i such that $u_i, v_i \in B_i$ for all i, and choose a T-block B such that $f(B_1, \ldots, B_n) \subseteq B$. According to item (2) of this theorem we have

$$u_i + v_i \sim_{B_i} v_i + u_i \tag{5.4}$$

for all i. Now consider the T, T-matrix

$$\begin{bmatrix} f(\mathbf{u} + \mathbf{u}) & f(\mathbf{u} + \mathbf{v}) \\ f(\mathbf{v} + \mathbf{u}) & f(\mathbf{v} + \mathbf{v}) \end{bmatrix}.$$

All entries lie in B, so from part (3) we get that

$$f(\mathbf{u} + \mathbf{u}) + f(\mathbf{v} + \mathbf{v}) \sim_B f(\mathbf{u} + \mathbf{v}) + f(\mathbf{v} + \mathbf{u})\,.$$

The left hand side is $f(\mathbf{u}) + f(\mathbf{v})$, since $+$ is idempotent. The right side is $\sim_B$-related to $f(\mathbf{u} + \mathbf{v})$ since, by (5.4), Claim 3.17 and the idempotence of $+$ we have

$$f(\mathbf{u} + \mathbf{v}) + f(\mathbf{v} + \mathbf{u}) \sim_B f(\mathbf{u} + \mathbf{v}) + f(\mathbf{u} + \mathbf{v}) = f(\mathbf{u} + \mathbf{v})\,.$$

Since $\sim_B$ is transitive on B we get $f(\mathbf{u}) + f(\mathbf{v}) \sim_B f(\mathbf{u} + \mathbf{v})$, hence $f(\mathbf{u}) + f(\mathbf{v}) \sim_T f(\mathbf{u} + \mathbf{v})$.

Now we prove (5) for polynomials. Consider the relation $\rho \subseteq A^3$ consisting of all triples (u, v, w) for which there is a T-block B such that $u, v, w \in B$ and $u + v \sim_B w$. It follows from the fact that item (5) is true when f is a basic operation, together with Claim 3.17 and the transitivity of $\sim_{B'}$ on B' for any block B', that ρ is compatible with the basic operations of $\mathbf{A}$. It is compatible with all constant operations on A because $+$ is idempotent. Thus ρ is compatible with all polynomial operations of $\mathbf{A}$. Since ρ contains all triples of the form $(u, v, u + v)$, it therefore contains $\big(p(\mathbf{u}), p(\mathbf{v}), p(\mathbf{u} + \mathbf{v})\big)$, establishing that $p(\mathbf{u} + \mathbf{v}) \sim_T p(\mathbf{u}) + p(\mathbf{v})$. □

We can prove more if T is rectangular.

THEOREM 5.14. *Let $\mathcal{V}$ be a variety with a join term $+$, and let T be a tolerance on an algebra $\mathbf{A} \in \mathcal{V}$. If $\sqsupseteq$ is a compatible partial order of $\mathbf{A}$ for which $\mathbf{R}(T, T; \sqsupseteq)$ holds, then the following are true.*

(1) *The relation* $\sim_T$ *is the equality relation on* $\mathbf{A}$.
(2) $\geq_T$ *equals* $\sqsupseteq \cap T$, *and also equals* $\{(u, v) \in T \mid u = u + v\}$.
(3) *If* $\begin{bmatrix} a & b \\ c & d \end{bmatrix} \in M(T, T)$, *then* $a + d = b + c$.
(4) *If* f *is a polynomial of* $\mathbf{A}$ *and* $\mathbf{u}\ T\ \mathbf{v}$, *then*

$$f(\mathbf{u} + \mathbf{v}) = f(\mathbf{u}) + f(\mathbf{v})\,.$$

(5) $+\colon T \to \mathbf{A}$ *is the unique compatible semilattice operation supported by* T.

PROOF. If (r, s) is a $(T, T; 0)$-pair, then there is a matrix

$$\begin{bmatrix} p & q \\ r & s \end{bmatrix} \in M(T, T)$$

with $q = r$. Since $r \sqsupseteq r$ and $r \sqsupseteq q\ (= r)$, we get $r \sqsupseteq s$ because $\mathbf{R}(T, T; \sqsupseteq)$ holds. Thus the relation $\sqsupseteq$ contains all $(T, T; 0)$-pairs, hence contains $\geq_T$. Therefore, the reflexive relation $\sim_T$ is contained in the intersection of $\sqsupseteq$ and its converse, which forces $\sim_T$ to be the equality relation. This proves item (1), and together with Theorem 5.13 (3) and (5) also proves items (3) and (4).

We will show that $\geq_T$ is contained in $\sqsupseteq \cap T$, which is contained in $\{(u, v) \in T \mid u = u + v\}$, which is contained in $\geq_T$. It follows from Definition 3.15 that $\geq_T$ is contained in T, and it follows from the argument of the previous paragraph that $\geq_T$ is contained in $\sqsupseteq$, so $\geq_T$ is contained in $\sqsupseteq \cap T$. Next, choose any pair (u, v) in $\sqsupseteq \cap T$. Since $\sqsupseteq$ is reflexive and compatible, and since $u \sqsupseteq v$, we get that $u = u + u \sqsupseteq u + v$. But $u + v \geq_T u$, according to the definition of a join term, so $u + v \sqsupseteq u$ by the argument of the first paragraph. This shows that both $(u, u + v)$ and $(u + v, u)$ belong to the partial order $\sqsupseteq$, so $u = u + v$. Hence $\sqsupseteq \cap T$ is contained in $\{(u, v) \in T \mid u = u + v\}$. Finally, choose any $(u, v) \in T$ such that $u = u + v$. If B is a block of T containing u and v, then $u = u + v \geq_B v$, since $\geq_B$ is the join order for $+$ on B (Theorem 5.13 (2)). Therefore $u \geq_T v$, proving that $\{(u, v) \in T \mid u = u + v\}$ is contained in $\geq_T$. This completes the proof of item (2).

For item (5), it follows from Theorem 5.13 (2) and part (1) of this theorem that $+$ is a semilattice operation on the blocks of T. The property from item (4) of this corollary, in the cases where f is a basic operation of $\mathbf{A}$, is exactly the property that $+\colon T \to \mathbf{A}$ is a homomorphism. Thus $+$ is a compatible semilattice operation on T. If $s\colon T \to \mathbf{A}$ is another compatible semilattice operation, then using the commutativity and idempotence of s and $+$, and the fact that $s\colon T \to \mathbf{A}$ is a homomorphism, and that $+$ is a term, we have for any $(u, v) \in T$ that

$$\begin{aligned} s(u, v) &= s(u, v) + s(v, u) \\ &= s(u + v, v + u) \\ &= s(u + v, u + v) \\ &= u + v\,. \end{aligned}$$

Thus $+$ is unique. □

COROLLARY 5.15. *Let* $\mathcal{V}$ *be a variety. The following are equivalent.*

(1) $\mathcal{V}$ *has no member with a nonzero congruence that is simultaneously rectangular and abelian.*
(2) $\mathcal{V}$ *has no member with a nonzero tolerance that is simultaneously rectangular and abelian.*
(3) $\mathcal{V}$ *satisfies a nontrivial idempotent Maltsev condition.*

PROOF. Since tolerances are congruences, (2) $\Longrightarrow$ (1).

A strongly abelian congruence is both rectangular and abelian, so if (1) holds, then $\mathcal{V}$ omits strongly abelian abelian congruences. Hence $\mathcal{V}$ satisfies a nontrivial idempotent Maltsev condition, according to Theorem 3.13, showing that (1) $\Longrightarrow$ (3).

Finally, if $\mathcal{V}$ satisfies a nontrivial idempotent Maltsev condition, then it has a join term, $+$. Suppose that $\mathbf{A} \in \mathcal{V}$ has a tolerance T that is rectangular and abelian. By Theorem 5.14 (5), $+$ is a semilattice operation on T-blocks. Therefore, if $(a, b) \in T$ and $a \geq b$ in the semilattice order, then

$$\begin{bmatrix} a & a \\ a & b \end{bmatrix} = \begin{bmatrix} a+a & a+b \\ b+a & b+b \end{bmatrix} \in M(T,T)\,.$$

But since T is also abelian, the equality in the first row of this matrix implies equality in the second row. Hence $T = 0$, and (3) $\Longrightarrow$ (2). □

One can use tame congruence theory, [**34**], to prove that a finite simple algebra has a nontrivial tolerance that is (a) simultaneously rectangular and abelian if and only if it has one that is (b) strongly rectangular if and only if it has one that is (c) strongly abelian. So perhaps it is not too surprising that the varieties that omit tolerances of type (a) are the same as those that omit type (b) or (c). However, (a), (b) and (c) are truly different properties of tolerances, even for finite algebras, as we explain in the next example.

EXAMPLE 5.16. This example is an extension of Example 3.2. As in that example, let X be a set and let $\mathcal{M} \subseteq X^{2\times 2}$ be a set of 2×2 matrices with entries in X. Define $\mathbf{A}(\mathcal{M})$ as in Example 3.2. Suppose that $\gg$ is a quasi-order defined on the set X such that whenever $M = \begin{bmatrix} p & q \\ r & s \end{bmatrix}$ belongs to $\mathcal{M}$, then

$$u \gg q \quad \& \quad u \gg r \quad \iff \quad u \gg s \quad \& \quad u \gg p$$

for any $u \in X$. Let $\sqsupseteq$ be the least extension of $\gg$ to a quasi-order on $A = X \cup Y \cup \{0\}$ (namely, $\sqsupseteq$ is the union of $\gg$ with the equality relation on A). Then $\sqsupseteq$ is a compatible quasi-order of $\mathbf{A}(\mathcal{M})$, and it is clear that $\mathbf{R}(R, S; \sqsupseteq)$ and $\mathbf{R}(T, T; \sqsupseteq)$ both hold (since for $u \neq v$ we have $u \sqsupseteq v$ if and only if $u \gg v$). Thus, we are again in the situation that we may ignore algebraic considerations and think set-theoretically when considering relationships between rectangulation and the term conditions introduced earlier.

To see, for example, that there is a finite algebra with a tolerance that is rectangular and abelian but not strongly rectangular (hence not strongly abelian), it is entirely sufficient to write down the single matrix $M = \begin{bmatrix} 1 & 3 \\ 3 & 2 \end{bmatrix}$. Then, for $\mathcal{M} = \{M\}$, and $\mathbf{A}(\mathcal{M})$ and T defined as in Example 3.2, the set of nontrivial T,T-matrices are just those that can be obtained from M by permuting its entries by some combination of row or column interchanges or transposes. As a T,T-matrix, M formally conflicts with $\mathbf{SR}(T, T; 0)$. But it does not formally conflict with $\mathbf{C}(T, T; 0)$ or $\mathbf{R}(T, T; \gg)$ where $\gg$ is the partial order on X generated by the pairs $(3, 1)$ and $(3, 2)$. Therefore, on the algebra $\mathbf{A}(\mathcal{M})$, the tolerance T is rectangular and abelian but not strongly rectangular.

The construction of Example 5.16 can be used to show that it is possible for the transitive closure of the $(T, T; 0)$-pairs to be a partial order even when T is not rectangular. For example, this occurs when

$$\mathcal{M} = \left\{ \begin{bmatrix} 1 & 3 \\ 3 & 2 \end{bmatrix}, \begin{bmatrix} 1 & 4 \\ 4 & 2 \end{bmatrix} \right\}.$$

Generalizations of this example show that the construction of the rectangular commutator of two tolerances can be arbitrarily complicated in general. The situation is better in varieties with join terms.

THEOREM 5.17. *Let $\mathcal{V}$ be a variety with a join term $+$, and let T be a tolerance on an algebra $\mathbf{A} \subset \mathcal{V}$. Let ρ be any congruence on $\mathbf{A}$ that contains $\sim_T$, and let $\mathbf{C} = \mathbf{A}/\rho$ and $S = T/\rho$.*

(1) *If $a_1 \geq_S a_2 \geq_S \cdots \geq_S a_n \geq_S a_1$ in $\mathbf{C}$, then $a_1 = a_2 = \cdots = a_n$.*
(2) *If $\sqsupseteq$ is the transitive closure of $\geq_S$, then $\sqsupseteq$ is a partial order of C whose restriction to any S-block agrees with $\geq_S$.*
(3) *$\mathbf{R}(S, S; \sqsupseteq)$ holds.*
(4) *The transitive closure of $\sim_T$ is $[T, T]_{\mathbf{R}}$.*

PROOF. First we prove (1). We may assume that all (a_i, a_{i+1}) (with the understanding that $n + 1 = 1$) are $(S, S; 0)$-pairs. Now choose $n \geq 2$ minimal among such cycles for which the statement is false. We argue that $a_1 + a_2 = a_2 + a_1 = a_1$. As (a_1, a_2) is an $(S, S; 0)$-pair, there exists a (T, T)-matrix

$$\begin{bmatrix} a & b \\ c & d \end{bmatrix}$$

such that $b/\rho = c/\rho = a_1$ and $d/\rho = a_2$. Choose a T-block B containing these four elements. By Theorem 5.13 we have that $b+c \sim_B a+d$, and that $+$ is a semilattice operation on B modulo $\sim_B$, which implies that $(b+c)+d \sim_B d+(b+c) \sim_B b+c$. The relation $\sim_B$ is contained in ρ, and therefore in $\mathbf{A}/\rho$ we have $(a_1 + a_1) + a_2 = a_2 + (a_1 + a_1) = a_1 + a_1$. But $+$ is idempotent, so $a_1 + a_2 = a_2 + a_1 = a_1$. A similar argument shows that $a_i + a_{i+1} = a_{i+1} + a_i = a_i$ for all i.

Consider

$$a_1 + a_2 \geq_S a_2 + a_3 \geq_S a_2 + a_4 \geq_S \cdots \geq_S a_2 + a_n \geq_S a_2 + a_1 .$$

This is a cycle of length $n - 1$. We show that every link in it is an $(S, S; 0)$-pair. Indeed, the sum of two (S, S)-matrices is also an (S, S)-matrix, hence the sum of two (S, S)-triples is an (S, S)-triple as well. Now $(a_1, a_1; a_2)$ and $(a_2, a_2; a_3)$ are (S, S)-triples, and so their sum $(a_1 + a_2, a_1 + a_2; a_2 + a_3)$ is an (S, S)-triple, proving that the first link is indeed an $(S, S; 0)$-pair. Similarly, $(a_2, a_2; a_2)$ and $(a_i, a_i; a_{i+1})$ are (S, S)-triples, and their sum shows that all the other links in the chain are $(S, S; 0)$-pairs, too.

The minimality of n then forces this chain to be trivial, that is, all links are equal. Thus $a_1 + a_2 = a_2 + a_3$. But we know that $a_1 + a_2 = a_1$ and $a_2 + a_3 = a_2$. Therefore $a_1 = a_2$. This contradicts the minimality of n in the original chain. This argument works for every $n > 1$, and so (1) is proved.

To prove (2) suppose that $u \sqsupseteq v$ and $v \sqsupseteq u$ hold. Then there is a $\geq_S$-chain from u to v, and from v to u. Thus the statement in (1) shows that $u = v$. Therefore $\geq$ is indeed a partial order. To prove the other statement, suppose that

$$b_1 \geq_S b_2 \geq_S b_3 \geq_S \cdots \geq_S b_n ,$$

where $(b_1, b_n) \in S$. Consider

$$b_1 + b_1 \geq_S b_2 + b_1 \geq_S b_3 + b_1 \geq_S \cdots \geq_S b_n + b_1 . \tag{5.5}$$

Here $b_n + b_1 \geq_S b_1 = b_1 + b_1$ by the properties of $+$, since b_n and b_1 are S-related. Hence (5.5) is a cycle, which must be trivial by (1). Therefore $b_1 = b_n + b_1 \geq_S b_n$, proving (2).

Now we prove that $\mathbf{R}(S, S; \sqsupseteq)$ holds. Suppose that

$$\begin{bmatrix} a & b \\ c & d \end{bmatrix} \in M(S, S),$$

and $u \in C$ satisfies that $u \sqsupseteq b$ and $u \sqsupseteq c$. If B is an S-block that contains these four elements, then Theorem 5.13 (3) guarantees that $b{+}c \sim_B a{+}d$. But the relation $\sim_B$ is the intersection of $\geq_B$ and its converse, the relation $\geq_B$ is contained in $\geq_S$, and $\geq_S$ is antisymmetric by item (2). Thus $\sim_B$ is equality on B, yielding $b + c = a + d$. As $+$ is idempotent and $\sqsupseteq$ is compatible, we get that $u = u + u \sqsupseteq b + c = a + d$. But $a + d \geq_S d$ by the definition of a join term, so $u \sqsupseteq a + d \sqsupseteq d$, proving that $\mathbf{R}(S, S; \sqsupseteq)$ holds.

For item (4), let ρ be the transitive closure of $\sim_T$ in $\mathbf{A}$. Since $\sim_T$ is a tolerance, its transitive closure ρ is a congruence. Now suppose that $\gg$ is the least compatible partial order on $\mathbf{A}$ such that $\mathbf{R}(T, T; \gg)$ holds. If

$$\begin{bmatrix} p & q \\ r & s \end{bmatrix} \in M(T, T),$$

and $q = r$, then $r \gg r$ and $r \gg q$ $(= r)$, so $r \gg s$. Thus $\gg$ contains all $(T, T; 0)$-pairs, and therefore contains $\geq_T$. According to Definition 5.3, $[T, T]_{\mathbf{R}}$ is the intersection of $\gg$ and its converse, hence $[T, T]_{\mathbf{R}}$ contains the intersection of $\geq_T$ and its converse, hence contains $\sim_T$. Since ρ is the least congruence containing $\sim_T$, we get $\rho \leq [T, T]_{\mathbf{R}}$.

On the other hand, according to Theorem 5.4, $[T, T]_{\mathbf{R}}$ is the least congruence δ for which T/δ is a rectangular tolerance of $\mathbf{A}/\delta$. By items (2) and (3) of this theorem, ρ is a congruence with this property. Therefore $[T, T]_{\mathbf{R}} = \rho$. □

The following consequence of Theorem 5.17 is worth stating separately.

COROLLARY 5.18. *Let $\mathcal{V}$ be a variety with a join term, and let T be a tolerance on some $\mathbf{A} \in \mathcal{V}$. Then T is rectangular if and only if $\sim_T$ is the equality relation.*

PROOF. If T is rectangular, then $\sim_T$ is the equality relation according to Theorem 5.14 (1). On the other hand, if $\sim_T$ is the equality relation, then Theorem 5.17 (4) guarantees that $[T, T]_{\mathbf{R}} = 0$. □

Corollary 5.18 implies the following characterization of rectangular tolerances in varieties with join terms, which is a refinement of Theorem 5.6.

COROLLARY 5.19. *Let $\mathcal{V}$ be a variety with a join term $+$, and let T be a tolerance on an algebra $\mathbf{A} \in \mathcal{V}$. Then T is rectangular if and only if $+$ is a compatible semilattice operation on T.*

PROOF. If T is rectangular, then $+$ is a compatible semilattice operation on T by Theorem 5.14 (5). Conversely, suppose that $+$ is a compatible semilattice operation on T. If (r, s) is a $(T, T; 0)$-pair, then there is a matrix

$$\begin{bmatrix} p & q \\ r & s \end{bmatrix} = \begin{bmatrix} f(\mathbf{a}, \mathbf{u}) & f(\mathbf{a}, \mathbf{v}) \\ f(\mathbf{b}, \mathbf{u}) & f(\mathbf{b}, \mathbf{v}) \end{bmatrix} \in M(T, T)$$

with $q = r$. In this situation

$$q + r = f(\mathbf{a} + \mathbf{b}, \mathbf{v} + \mathbf{u}) = p + s\,,$$

and all elements p, q, r, s, $q + r$ and $p + s$ lie in a single T-block. Therefore $r = q + r = p + s \geq s$ in the semilattice order on that block. This means that the equation $r + s = r$ holds, which is a property that is independent of the block that we are considering. Thus if (r, s) is a $(T, T; 0)$-pair, then $r \geq s$ in the semilattice order on any block containing r and s. This conclusion implies that on any T-block the relation $\geq_T$ is contained in the semilattice order, and therefore that $\sim_T$ is the equality relation. Hence the tolerance T is rectangular by Corollary 5.18. □

The following corollary will be used in Section 9.2.

COROLLARY 5.20. *Let $\mathcal{V}$ be a variety with a join term $+$, and let T be a rectangular tolerance on an algebra $\mathbf{A} \in \mathcal{V}$. Assume that a and b are distinct T-related elements of $\mathbf{A}$. If ρ is a congruence on $\mathbf{A}$ that is maximal with respect to $(a, b) \notin \rho$, then T/ρ is a nontrivial rectangular tolerance on the subdirectly irreducible algebra $\mathbf{A}/\rho$.*

PROOF. If T is rectangular, then $\sim_T$ is the equality relation according to Theorem 5.14 (1). Therefore every congruence ρ on $\mathbf{A}$ contains $\sim_T$. From Theorem 5.17, T/ρ is a rectangular congruence of $\mathbf{A}/\rho$ for any ρ. Choosing ρ as described in the statement of this corollary guarantees that T/ρ is nontrivial and that $\mathbf{A}/\rho$ is subdirectly irreducible. □

5.3. Varieties Omitting Rectangular Tolerances

The purposes of this section are to show that a variety omits rectangular tolerances if and only if it omits rectangular congruences, and then to give three different Maltsev characterizations of the class of varieties with these properties.

LEMMA 5.21. *Let $\mathcal{V}$ be a variety with a join term and $\mathbf{F} = \mathbf{F}_{\mathcal{V}}(x, y)$ the free $\mathcal{V}$-algebra generated by the set $\{x, y\}$. Let $T = \mathrm{Tg}^{\mathbf{F}}(x, y)$, let $\theta = \mathrm{Cg}^{\mathbf{F}}(T)$, and let $B = x/\theta$. Then the relations $\geq_\theta$ and $\geq_T$ have the same restriction to B.*

PROOF. It follows from Lemma 3.16 (4) that $(\geq_T)|_B = \mathrm{tr.cl.}(S|_B)$, where S is the set of $(T, T; 0)$-pairs of $\mathbf{F}$, and that $(\geq_\theta)|_B = \mathrm{tr.cl.}(\overline{S}|_B)$, where $\overline{S}$ is the (larger) set of $(\theta, \theta; 0)$-pairs of $\mathbf{F}$. Thus, it is enough to show that if $(r, s) \in B^2$ is a $(\theta, \theta; 0)$-pair, then (r, s) is in the transitive closure of the set of $(T, T; 0)$-pairs. So choose a θ, θ-matrix of the form

$$\begin{bmatrix} p & r \\ r & s \end{bmatrix} = \begin{bmatrix} t(\mathbf{a}, \mathbf{c}) & t(\mathbf{a}, \mathbf{d}) \\ t(\mathbf{b}, \mathbf{c}) & t(\mathbf{b}, \mathbf{d}) \end{bmatrix}.$$

Let $h\colon \mathbf{F} \to \mathbf{F}$ be the endomorphism determined by $x \mapsto x$, $y \mapsto x$. The conditions $\mathbf{a} \equiv_\theta \mathbf{b}$ and $\mathbf{c} \equiv_\theta \mathbf{d}$ are equivalent to $h(\mathbf{a}) = h(\mathbf{b})$ and $h(\mathbf{c}) = h(\mathbf{d})$. As in the proof of Lemma 3.7, consider the following 4×4 matrix:

$$\begin{bmatrix} t(\mathbf{a}, \mathbf{c}) & t(\mathbf{a}, h(\mathbf{c})) & = & t(\mathbf{a}, h(\mathbf{d})) & t(\mathbf{a}, \mathbf{d}) \\ t(h(\mathbf{a}), \mathbf{c}) & t(h(\mathbf{a}), h(\mathbf{c})) & = & t(h(\mathbf{a}), h(\mathbf{d})) & t(h(\mathbf{a}), \mathbf{d}) \\ \| & \| & & \| & \| \\ t(h(\mathbf{b}), \mathbf{c}) & t(h(\mathbf{b}), h(\mathbf{c})) & = & t(h(\mathbf{b}), h(\mathbf{d})) & t(h(\mathbf{b}), \mathbf{d}) \\ t(\mathbf{b}, \mathbf{c}) & t(\mathbf{b}, h(\mathbf{c})) & = & t(\mathbf{b}, h(\mathbf{d})) & t(\mathbf{b}, \mathbf{d}) \end{bmatrix}.$$

Recall from that earlier proof that the four elements in the middle are all equal to x and that the four 2×2 matrices in the corners are T, T-matrices. All elements of this 4×4 matrix are contained in B, because all elements are θ-related, $r, s \in B$ by assumption, and B is a block of θ.

Proceeding as we did in the proof of Lemma 3.7, let $k\colon \mathbf{F} \to \mathbf{F}$ be the endomorphism of $\mathbf{F}$ that sends x to y and y to y. Consider the analogous 4×4 matrix with respect to k, and apply $r(x, y)$ to these two matrices, the first matrix in the first argument, the second matrix in the second argument. As before, we get a 4×4 matrix which, after deleting one of the doubled middle columns and one of the doubled middle rows, is a 3×3 matrix of the form:

$$J = \begin{bmatrix} p & i & r \\ j & r & \ell \\ r & m & s \end{bmatrix}.$$

The four 2×2 matrices in the corners of this 3×3 matrix J are all T, T-matrices. From the matrix in the upper right corner we get that $r \geq_T \ell$. From the matrix in the lower left corner we obtain that $r \geq_T m$. From the matrix in the lower right corner we get, using Theorem 5.13 (4), that $\ell + m \geq_T s$. But $+$ is idempotent, $\geq_T$ is compatible, and $\geq_T$ is transitive on B, so $r = r + r \geq_T \ell + m \geq_T s$. □

THEOREM 5.22. *Let $\mathcal{V}$ be a variety. The following are equivalent.*

(1) *No member of $\mathcal{V}$ has a nontrivial rectangular congruence.*
(2) *No member of $\mathcal{V}$ has a nontrivial rectangular tolerance.*

PROOF. Clearly (2) $\Longrightarrow$ (1), since congruences are tolerances. The same type of argument as that used in the proof of Lemma 3.5 shows that to prove the converse it is enough to show that

$$(x, y) \in [\theta, \theta]_{\mathbf{R}} \implies (x, y) \in [T, T]_{\mathbf{R}}$$

where $\theta = \mathrm{Cg}^{\mathbf{F}}(x, y)$ and $T = \mathrm{Tg}^{\mathbf{F}}(x, y)$ are the congruence and tolerance generated by (x, y) in the free algebra $\mathbf{F} = \mathbf{F}_{\mathcal{V}}(x, y)$. So assume that $(x, y) \in [\theta, \theta]_{\mathbf{R}}$. This assumption, which is equivalent to the assumption that $\mathcal{V}$ has no member with a rectangular congruence, implies that $\mathcal{V}$ has no algebra with a strongly rectangular congruence. Theorems 3.21 and 3.23 now guarantee that $\mathcal{V}$ has a join term. This puts us into position to use Theorem 5.17. By part (4) of that theorem, the condition that $(x, y) \in [\theta, \theta]_{\mathbf{R}}$ means that (x, y) lies in the transitive closure of $\sim_\theta$. But according to Lemma 3.16 (5), the fact that θ is a congruence implies that $\sim_\theta$ is transitive. Therefore $x \sim_\theta y$. This implies that $x \geq_\theta y$ and $y \geq_\theta x$, so by Lemma 5.21 we get that $x \geq_T y$ and $y \geq_T x$. Since $B = x/\theta$ is the unique T-block containing x and y we have $x \geq_B y$ and $y \geq_B x$; i.e., $x \sim_B y$ and hence $x \sim_T y$. Theorem 5.17 (4) now guarantees that $(x, y) \in [T, T]_{\mathbf{R}}$. □

The proof of Theorem 5.22 shows that a variety $\mathcal{V}$ contains no member with a rectangular tolerance or congruence if and only if $x \sim_T y$ in $\mathbf{F} = \mathbf{F}_{\mathcal{V}}(x, y)$ for $T = \mathrm{Tg}^{\mathbf{F}}(x, y)$. This property is easily seen to be expressible by an idempotent Maltsev condition, which we record here.

THEOREM 5.23. *Let $\mathcal{V}$ be a variety. The following are equivalent.*

(1) *No member of $\mathcal{V}$ has a nontrivial rectangular tolerance.*
(2) *$x \sim_T y$ in $\mathbf{F} = \mathbf{F}_{\mathcal{V}}(x, y)$ for $T = \mathrm{Tg}^{\mathbf{F}}(x, y)$.*

(3) *There exists an* $m \geq 1$, *and sixary terms* $f_1, \ldots, f_m$ *such that* $\mathcal{V}$ *satisfies the following identities:*
  (i) $x \approx f_1(y, y, x, x, x, y)$,
  (ii) $f_m(y, x, x, x, y, y) \approx y$,
  (iii) $f_i(x, x, x, y, y, y) \approx f_i(y, y, x, x, x, y)$, $1 \leq i \leq m$,
  (iv) $f_i(y, x, x, x, y, y) \approx f_{i+1}(y, y, x, x, x, y)$, $1 \leq i \leq m-1$.

(4) *For every algebra* $\mathbf{A} \in \mathcal{V}$, *tolerance* S *on* $\mathbf{A}$, *and* $(a, b) \in S$, $a \geq_S b$ *holds.*

(5) *For every algebra* $\mathbf{A} \in \mathcal{V}$, *tolerance* S *on* $\mathbf{A}$, *and* $(a, b) \in S$, $a \sim_S b$ *holds.*

PROOF. The proof of Theorem 5.22 shows that (1) and (2) are equivalent.

If (2) holds, then $x \geq_T y$ in $\mathbf{F}$, so according to Lemma 3.20 there exists a sequence $f_1, \ldots, f_m$ of sixary terms satisfying the identities in (3).

Now suppose that (3) holds. Choose $\mathbf{A} \in \mathcal{V}$, a tolerance S on $\mathbf{A}$, an S-block B and $a, b \in B$. Since B is a block, it is closed under all the idempotent terms that appear in item (3). By substituting a and b for x and y in the identities of (3) we obtain elements of S witnessing that $a \geq_B b$. Namely, the fact that $(a, b) \in S$ implies that for each i the matrix

$$\begin{bmatrix} p_i & q_i \\ r_i & s_i \end{bmatrix} = \begin{bmatrix} f_i(a,b,a,b,a,b) & f_i(a,a,a,b,b,b) \\ f_i(b,b,a,a,a,b) & f_i(b,a,a,a,b,b) \end{bmatrix}$$

is in $M(S, S)$. Identities of type (iii) imply that $q_i = r_i$, so each (r_i, s_i) is an $(S, S; 0)$ pair. The other identities imply that the sequence of these pairs is a connected chain from a to b. Thus $a \geq_S b$, and item (4) is established.

The argument of the previous paragraph showed that $a \geq_B b$ for some S-block B, and a similar one shows that $b \geq_B a$ for the same block B. Hence $a \sim_B b$, which implies that $a \sim_S b$. This establishes item (5). Finally, item (2) is a special case of (5). □

Our next task will be to connect the foregoing with the Hobby–McKenzie term. First we need a lemma.

LEMMA 5.24. *If* $\mathbf{C}$ *is a nontrivial idempotent algebra with a compatible semilattice term, then there is an algebra* $\mathbf{D}$ *that is a homomorphic image of a subalgebra of* $\mathbf{C}$ *and is term equivalent to the 2-element semilattice.*

PROOF. Denote the semilattice operation by $+$, and call it join. Since the semilattice order is equationally definable by

$$x \geq y \iff x = x + y,$$

and $+$ is compatible, it follows that the semilattice order is a compatible relation. As C is not a singleton, there exist two distinct elements $0 < 1$ of $\mathbf{C}$. If $c_1, \ldots, c_k$ belong to the interval $[0, 1]$ (with respect to the semilattice ordering), and f is a term operation of $\mathbf{C}$, then

$$1 = f(1, \ldots, 1) \geq f(c_1, \ldots, c_k) \geq f(0, \ldots, 0) = 0$$

by the idempotence of f and the compatibility of $\geq$. Thus $[0, 1]$ is a nontrivial subalgebra of $\mathbf{C}$. Replacing $\mathbf{C}$ by this subalgebra, we may henceforth assume that $\mathbf{C}$ has least and largest elements 0 and 1.

If f is a k-ary term operation of $\mathbf{C}$, then since $+$ is compatible, the operation f must commute with $x_1 + \cdots + x_k$ on the matrix

$$\begin{bmatrix} x_1 & 0 & \cdots & 0 \\ 0 & x_2 & \cdots & 0 \\ \vdots & \vdots & & \vdots \\ 0 & 0 & \cdots & x_k \end{bmatrix}.$$

This says exactly that $f(x_1, \ldots, x_k) = f_1(x_1) + \cdots + f_k(x_k)$ where $f_i(x_i) := f(0, \ldots, 0, x_i, 0, \ldots, 0)$, with x_i in the i-th position. As f is idempotent, we have that

$$f_1(x) + \cdots + f_n(x) = x$$

for every x. In particular, $f_i(x) \leq x$ for every x. The functions f_i are of course endomorphisms for $+$, too, since f is compatible with $+$ and $0 + 0 = 0$.

Choose a maximal semilattice ideal I of $\mathbf{C}$ that does not contain 1. Let $F = C - I$ be the complementary filter. Let θ be the equivalence relation on C whose classes are I and F. We claim that θ is a congruence of $\mathbf{C}$, and that $\mathbf{C}/\theta$ is term equivalent to a semilattice. To show this, represent an arbitrarily chosen term f as a sum of unary polynomials as described in the previous paragraph. The condition $f_i(x) \leq x$ implies that $f_i(I) \subseteq I$. We prove that either $f_i(F) \subseteq F$, or $f_i(F) \subseteq I$. Assume that $f_i(F) \nsubseteq F$, and choose $u \in F$ so that $f_i(u) \in I$. By the maximality of I, there exists an element $a \in I$ such that $a + u = 1$. Since f_i is a $+$-endomorphism we have $f_i(1) = f_i(a) + f_i(u) \in I$. This shows that if $f_i(u) \in I$, then $f_i(1) \in I$, hence $f_i(C) \subseteq I$. Consequently each f_i preserves θ. Since $+$ also preserves θ, it follows that f preserves it, so θ is a congruence of $\mathbf{C}$.

The quotient $\mathbf{D} = \mathbf{C}/\theta$ is a 2-element idempotent algebra with a compatible semilattice term $+$. Any such algebra is term equivalent to the 2-element semilattice, as we now show. Denote the elements of D by 0 and 1 such that $0 < 1$ in the $+$-order. As above, each term operation of $\mathbf{D}$ may be represented as $f(x_1, \ldots, x_k) = f_1(x_1) + \cdots + f_k(x_k)$ where each f_i is a decreasing $+$-endomorphism. But since $D = \{0, 1\}$ this forces each f_i to be the identity function or the constant function with value 0. If $J \subseteq \{1, \ldots, k\}$ is the set of indices j such that f_j is the identity, then $J \neq \emptyset$, since f is idempotent, and $f(x_1, \ldots, x_k) = \sum_J x_j$. Hence f equals a semilattice term operation. □

Now we are in position to give a second Maltsev characterization of the class of varieties that omit rectangular tolerances.

THEOREM 5.25. *Let $\mathcal{V}$ be a variety. The following are equivalent.*

(1) *No member of $\mathcal{V}$ has a nontrivial rectangular tolerance.*
(2) *$\mathcal{V}$ satisfies an idempotent (linear) Maltsev condition that fails in the variety of semilattices.*
(3) *$\mathcal{V}$ has a Hobby–McKenzie term.*

PROOF. The equivalence of (2) and (3) is Theorem 2.16. To show that (1) $\Longrightarrow$ (2) we argue that the idempotent linear Maltsev condition from Theorem 5.23 (3) fails in the variety of semilattices. We give two proofs of this.

First, the Maltsev condition from Theorem 5.23 holds only in varieties omitting rectangular congruences. A calculation shows that the largest congruence on the 2-element semilattice is rectangular. Therefore the Maltsev conditon fails in the variety of semilattices.

For the second proof, observe that if $f_1(x_1, \ldots, x_6)$ is a semilattice term, then identity (i) of Theorem 5.23 (3), $x \approx f_1(y, y, x, x, x, y)$, implies that f_1 does not involve the variables x_1, x_2 or x_6. (Otherwise, by putting the absorbing element of the 2-element semilattice in for these variables and the nonabsorbing element for the other variables one obtains the contradiction that these two elements are equal.) This, together with the first instance of identity (iii), $f_1(x, x, x, y, y, y) \approx f_1(y, y, x, x, x, y)$ ($\approx x$), shows that f_1 does not involve x_4, x_5 or x_6 either. Therefore, by idempotence, $f_1(x_1, \ldots, x_6) \approx x_3$. This argument propagates down the chain of f_i's via identity (iv), showing that each term satisfies $f_i(x_1, \ldots, x_6) \approx x_3$. Identity (ii) then reduces to $x \approx y$, which fails in the variety of semilattices.

Now we prove (2) $\Longrightarrow$ (1). Assume that $\mathcal{V}$ satisfies a (not necessarily linear) idempotent Maltsev condition, called W, that fails in the variety of semilattices. Let $T = \mathrm{Tg}(x, y)$ be the tolerance generated by (x, y) on the free algebra $\mathbf{F} = \mathbf{F}_{\mathcal{V}}(x, y)$, and let B be the T-block of idempotent binary terms. The set B is closed under all idempotent terms of $\mathcal{V}$, so let $\mathbf{B}$ be the algebra with universe B and basic operations all idempotent terms of $\mathcal{V}$. Then $\mathbf{B}$ is an idempotent algebra that satisfies the idempotent Maltsev condition W. The relation $\sim_T$ restricted to B is a congruence θ of $\mathbf{B}$.

Since the Maltsev condition W is nontrivial, the variety generated by $\mathbf{B}$ has a join term $+$. By Theorem 5.13 (2) and (5), $+$ is a compatible semilattice term of the idempotent algebra $\mathbf{C} = \mathbf{B}/\theta$. If $\mathbf{C}$ is nontrivial, then Lemma 5.24 guarantees that some homomorphic image $\mathbf{D}$ of some subalgebra $\mathbf{C}$ is term equivalent to the 2-element semilattice. The Maltsev condition W is satisfied by $\mathbf{C}$, hence by $\mathbf{D}$, hence by the variety of semilattices. But we have assumed the contrary, so we conclude that $\mathbf{C}$ is trivial. This proves that $\mathbf{B}$ is a single $\sim_T$ class, and therefore that $x \sim_T y$ in $\mathbf{F}_{\mathcal{V}}(x, y)$. According to Theorem 5.23, this means that $\mathcal{V}$ contains no algebra with a nontrivial rectangular tolerance. □

In the preceding proof the implication (3) $\Longrightarrow$ (1) is verified indirectly. Before proceeding, we shall describe a direct proof of this implication. Suppose that $f(x_1, \ldots, x_n)$ is a Hobby–McKenzie term for $\mathcal{V}$. Recall from Chapter 2 that for a subset $U \subseteq N := \{1, \ldots, n\}$ the expression $f_U(x, y)$ denotes the term obtained from f by substituting x for x_i if $i \in U$ and y for x_i otherwise. The statement that f is a Hobby–McKenzie term for $\mathcal{V}$ is precisely the statement that f is idempotent and for any nonempty subset $U \subseteq N$ there exist subsets $V, W \subseteq N$ with $U \subseteq V$, $U \not\subseteq W$ and $\mathcal{V} \models f_V(x, y) \approx f_W(x, y)$.

As usual, let $\mathbf{F} = \mathbf{F}_{\mathcal{V}}(x, y)$, $T = \mathrm{Tg}^{\mathbf{F}}(x, y)$, and B the T-block of idempotent binary terms. Our aim is to prove that $x \sim_T y$ using the equations of the Hobby–McKenzie term. Recall that the restriction of $\sim_T$ to B is transitive. Let $U \subseteq N$ be minimal under inclusion with respect to the property that $f_U(x, x + y) \sim_T x$. (Such a set U exists, since $f_N(x, x + y) = x \sim_T x$.) If U is not empty, then there exist $V, W \subseteq N$ such that $U \subseteq V$ but $U \not\subseteq W$, and $f_V(x, x + y) = f_W(x, x + y)$. Hence by Theorem 5.13,

$$f_U(x, x + y) + f_W(x, x + y) \sim_T f_{U \cap W}(x, x + y)\,, \tag{5.6}$$

and

$$f_U(x, x + y) + f_V(x, x + y) \sim_T f_{U \cap V}(x, x + y) = f_U(x, x + y)\,. \tag{5.7}$$

Since $f_V(x, x+y) = f_W(x, x+y)$, the left hand sides of lines (5.6) and (5.7) are the same. Therefore

$$f_{U \cap W}(x, x+y) \sim_T f_U(x, x+y) \sim_T x\,.$$

This contradicts the minimality of U unless $U = \emptyset$, which leads to the conclusion that $x \sim_T f_\emptyset(x, x+y) = x+y$. A symmetric argument shows that $y \sim_T x+y$, and therefore $x \sim_T y$. At this point, Theorem 5.23 may be invoked to deduce that no algebra in $\mathcal{V}$ has a rectangular tolerance.

We turn to our third Maltsev characterization of the class of varieties that omit rectangular tolerances. This new Maltsev condition involves four variable terms, and somewhat resembles Alan Day's Maltsev condition characterizing the class of congruence modular varieties (cf. [**8**]).

Definition 5.26. Let $\mathbf{A}$ be an algebra. If $\mathbf{Con}(\mathbf{A})$ has a sublattice isomorphic to the lattice $\mathbf{D}_2$ with $(\theta \circ \mu) \cap (\nu \circ \delta) \nsubseteq \alpha$, then we call this sublattice a **special**

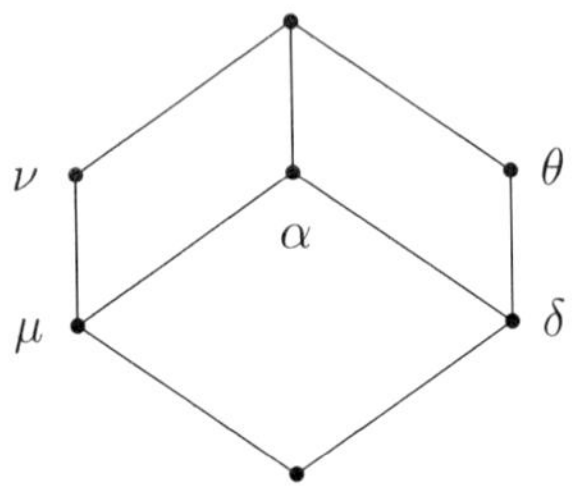

Figure 5.1. $\mathbf{D}_2$

$\mathbf{D}_2$ in $\mathbf{Con}(\mathbf{A})$. We say that $\mathcal{V}$ **omits special $\mathbf{D}_2$'s** if for no $\mathbf{A} \in \mathcal{V}$ does $\mathbf{Con}(\mathbf{A})$ have a sublattice that is a special $\mathbf{D}_2$.

As an aid to recognizing copies of $\mathbf{D}_2$ in congruence lattices, we record the following.

Lemma 5.27. *A presentation of $\mathbf{D}_2$ relative to the variety of all lattices is $\langle G \mid R\rangle$ where $G = \{x, y, z\}$ and R consists of the relations:*

(i) $x \wedge z \leq y$,
(ii) $x \leq z \vee (y \wedge x)$,
(iii) $z \leq x \vee (y \wedge z)$, *and*
(iv) $y \leq (x \wedge y) \vee (y \wedge z)$.

Moreover any lattice generated by G and satisfying the relations in R and also satisfying $x \nleq y$ is isomorphic to $\mathbf{D}_2$.

Proof. The reader can easily verify that the copy of $\mathbf{D}_2$ in Figure 5.2 is generated by x, y and z, and that the relations in the statement of the lemma hold for these elements. To show that it is a presentation of $\mathbf{D}_2$ we must prove from the relations (and the laws of lattice theory) that the seven elements x, y, z, $p := x \wedge y$, $q := z \wedge y$, $s := x \wedge y \wedge z$, and $t := x \vee y \vee z$ are closed under meet and join.

It is clear that the seven elements are ordered as they should be, and consequently s and t are the bottom and top elements of the sublattice they generate. The only incomparable doubletons are $\{x, y\}$, $\{x, z\}$, $\{x, q\}$, $\{y, z\}$, $\{p, z\}$ and $\{p, q\}$.

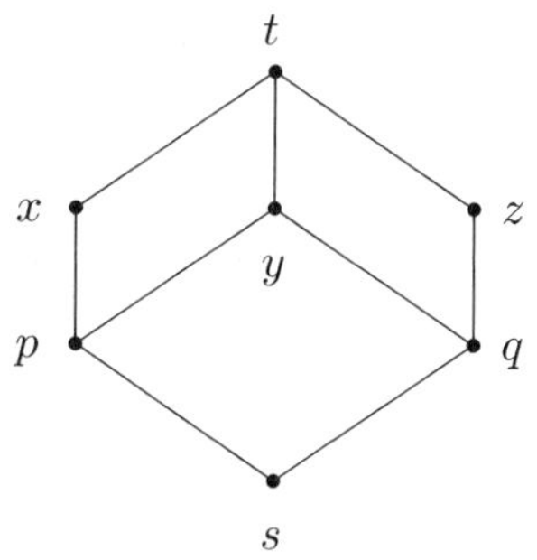

FIGURE 5.2

By symmetry, we do not need to consider the meets or joins of $\{y, z\}$ or $\{p, z\}$. For the remaining cases:

- $\{x, y\}$: $x \wedge y = p$ by the definition of p. By relation (iii), $x \vee y$ is above z. Since $x \vee y$ is formally below $x \vee y \vee z = t$, it must be equal to it.
- $\{x, z\}$: $x \wedge z$ is below y by relation (i), so $x \wedge z = x \wedge y \wedge z = s$, as desired. $x \vee z$ is above $(x \wedge y) \vee (y \wedge z)$, since corresponding joinands are larger, and $(x \wedge y) \vee (y \wedge z) \geq y$ by (iv). Therefore $x \vee z = x \vee y \vee z = t$.
- $\{x, q\}$: $x \wedge q = x \wedge (y \wedge z) = s$ holds by the definition of q. $x \vee q = x \vee (y \wedge z)$ is above z by (iii), therefore it is above $x \vee z = t$ and hence equal to t.
- $\{p, q\}$: $p \wedge q$ is below $x \wedge z$, which has already been shown to be the bottom element, so $p \wedge q = s$. By (iv) we have $y \leq (x \wedge y) \vee (y \vee z) = p \vee q \leq y$, so $p \vee q = y$.

This proves that $\langle G \mid R \rangle$ presents $\mathbf{D}_2$. For the last claim of the lemma, any lattice generated by G satisfying the relations R which is not isomorphic to $\mathbf{D}_2$ must be a proper quotient of $\mathbf{D}_2$. Since $\mathbf{D}_2$ is a subdirectly irreducible lattice whose least congruence collapses p and x, any such lattice will satisfy $x = p \leq y$. Therefore, if a lattice is generated by G and satisfies R but does not satisfy $x \leq y$, then the lattice is isomorphic to $\mathbf{D}_2$. □

THEOREM 5.28. *The following are equivalent for a variety $\mathcal{V}$.*

(1) *$\mathcal{V}$ satisfies an idempotent Maltsev condition that fails in the variety of semilattices.*

(2) *Whenever θ, μ, ν and δ are congruences on some $\mathbf{A} \in \mathcal{V}$, then $(\theta \circ \mu) \cap (\nu \circ \delta) \subseteq [(\theta \vee \delta) \wedge (\nu \vee \mu)] \vee \delta \vee \mu$.*

(3) *$\mathcal{V}$ has a sequence of terms $f_i(x, y, u, v)$ for $0 \leq i \leq 2m + 1$, such that*

  (i) $\mathcal{V} \models f_0(x, y, u, v) \approx x$;

  (ii) $\mathcal{V} \models f_i(x, y, y, y) \approx f_{i+1}(x, y, y, y)$ *for even i;*

  (iii) $\mathcal{V} \models f_i(x, x, y, y) \approx f_{i+1}(x, x, y, y)$ *and*
  $\mathcal{V} \models f_i(x, y, x, y) \approx f_{i+1}(x, y, x, y)$ *for odd i;*

  (iv) $\mathcal{V} \models f_{2m+1}(x, y, u, v) \approx v$.

(4) *$\mathcal{V}$ omits special $\mathbf{D}_2$'s.*

PROOF. We first establish the equivalence of (2), (3) and (4), and connect them to item (1) afterward.

Assume that (2) holds. On the free algebra $\mathbf{F} = \mathbf{F}_{\mathcal{V}}(x, y, u, v)$ let $\theta = \mathrm{Cg}^{\mathbf{F}}(x, u)$, $\mu = \mathrm{Cg}^{\mathbf{F}}(u, v)$, $\nu = \mathrm{Cg}^{\mathbf{F}}(x, y)$, and $\delta = \mathrm{Cg}^{\mathbf{F}}(y, v)$. Since $(x, v) \in (\theta \circ \mu) \cap (\nu \circ \delta)$, and (2) holds:

$$(\theta \circ \mu) \cap (\nu \circ \delta) \subseteq [(\theta \vee \delta) \wedge (\nu \vee \mu)] \vee \delta \vee \mu\,,$$

we have $(x, v) \in [(\theta \vee \delta) \wedge (\nu \vee \mu)] \vee \delta \vee \mu$. Therefore, there is a sequence $f_0, f_1, \ldots,$ f_{2m+1} of elements of $\mathbf{F}$ such that $f_0 = x$, $f_{2m+1} = v$, $(f_i, f_{i+1}) \in \delta \vee \mu$ for even i while $(f_i, f_{i+1}) \in (\theta \vee \delta) \wedge (\nu \vee \mu)$ for odd i. If one chooses 4–ary terms representing the f_i's, it is straightforward to see that the equations of (3) hold.

Assume that (3) holds and that some $\mathbf{A} \in \mathcal{V}$ has congruences μ, ν, α, δ and θ that generate a sublattice of $\mathbf{Con}(\mathbf{A})$ isomorphic to $\mathbf{D}_2$ labeled as in Figure 5.1. Choose any $(a, d) \in (\theta \circ \mu) \cap (\nu \circ \delta)$. There exist elements $b, c \in A$ such that $a \equiv_\nu b \equiv_\delta d$ and $a \equiv_\theta c \equiv_\mu d$. The pairs $\big(f_i(a, b, c, d), f_{i+1}(a, b, c, d)\big)$ belong to $\delta \vee \mu = \alpha$ for even i and $(\theta \vee \delta) \wedge (\nu \vee \mu) \leq \alpha$ for odd i. Hence $a = f_0(a, b, c, d)$ is α-related to $f_{2m+1}(a, b, c, d) = d$. Since $(a, d) \in (\theta \circ \mu) \cap (\nu \circ \delta)$ was chosen arbitrarily, this implies that $(\theta \circ \mu) \cap (\nu \circ \delta) \subseteq \alpha$. Therefore this copy of $\mathbf{D}_2$ is not a special $\mathbf{D}_2$. The argument works for every $\mathbf{D}_2$ that appears in a congruence lattice of a member of $\mathcal{V}$, so $\mathcal{V}$ omits special $\mathbf{D}_2$'s and item (4) holds.

Next we argue the contrapositive of (4) $\Longrightarrow$ (2), so assume that some $\mathbf{A} \in \mathcal{V}$ has congruences μ, ν, δ and θ such that

$$(\theta \circ \mu) \cap (\nu \circ \delta) \not\subseteq [(\theta \vee \delta) \wedge (\nu \vee \mu)] \vee \delta \vee \mu\,.$$

Choose $(a, d) \in (\theta \circ \mu) \cap (\nu \circ \delta)$ satisfying

$$(a, d) \notin [(\theta \vee \delta) \wedge (\nu \vee \mu)] \vee \delta \vee \mu\,,$$

as well as elements $b, c \in A$ such that $a \equiv_\nu b \equiv_\delta d$ and $a \equiv_\theta c \equiv_\mu d$. Let $\theta^* := \mathrm{Cg}^{\mathbf{A}}\big((a, c), (b, d)\big)$, $\nu^* := \mathrm{Cg}^{\mathbf{A}}\big((a, b), (c, d)\big)$, and

$$\alpha^* := \mathrm{Cg}^{\mathbf{A}}\big((b, d), (c, d)\big) \vee (\theta^* \wedge \nu^*)\,.$$

We will show that the congruences $x := \theta^*$, $y := \alpha^*$, and $z := \nu^*$ satisfy the presentation for $\mathbf{D}_2$ given in Lemma 5.27.

Item (i) of Lemma 5.27 requires that $\theta^* \wedge \nu^* \leq \alpha^*$, which is an immediate consequence of the definition of α^*.

Item (ii) requires that $\theta^* \leq \nu^* \vee (\alpha^* \wedge \theta^*)$. Since ν^* contains (a, b) and (c, d) while $\alpha^* \wedge \theta^*$ contains (b, d), the join is the congruence generated by $X \times X$ where $X = \{a, b, c, d\}$. Hence the join contains the generators (a, c) and (b, d) of θ^*. Item (iii) is symmetric to item (ii).

For item (iv), the congruence $\theta^* \wedge \alpha^*$ contains $\{(b, d)\} \cup (\theta^* \wedge \nu^*)$, since this is true of each of θ^* and α^*. Similarly $\nu^* \wedge \alpha^*$ contains $\{(c, d)\} \cup (\theta^* \wedge \nu^*)$. This shows that $(\theta^* \wedge \alpha^*) \vee (\alpha^* \wedge \nu^*)$ contains all the generators of α^*. Hence $\alpha^* \leq (\theta^* \wedge \alpha^*) \vee (\alpha^* \wedge \nu^*)$.

Finally we must show that $\theta^* \not\leq \alpha^*$. Assume that this is not the case. Since $(a, c) \in \theta^*$, this assumption forces $(a, c) \in \alpha^*$. By definition, $\theta^* \leq (\theta \vee \delta)$, $\nu^* \leq (\nu \vee \mu)$, $\mathrm{Cg}^{\mathbf{A}}(b, d) \leq \delta$ and $\mathrm{Cg}^{\mathbf{A}}(c, d) \leq \mu$, so

$$\alpha^* = (\theta^* \wedge \nu^*) \vee \mathrm{Cg}^{\mathbf{A}}(c, d) \vee \mathrm{Cg}^{\mathbf{A}}(b, d) \leq [(\theta \vee \delta) \wedge (\nu \vee \mu)] \vee \delta \vee \mu\,.$$

Combining the facts that $(a, c) \in \alpha^*$ and $(c, d) \in \mu$, we obtain that the right hand side of the previous displayed line contains (a, d). This is contrary to the choice of (a, d), so we are forced to conclude that $\theta^* \not\leq \alpha^*$. This completes the proof that

$\{\theta^*, \alpha^*, \nu^*\}$ generates a copy of $\mathbf{D}_2$ in $\mathbf{Con}(\mathbf{A})$. Indeed, this is a *special* $\mathbf{D}_2$ since $a \equiv_{\nu^*} b \equiv_{\alpha^* \wedge \theta^*} d$ and $a \equiv_{\theta^*} c \equiv_{\alpha^* \wedge \nu^*} d$, so

$$(a, d) \in \big(\theta^* \circ (\alpha^* \wedge \nu^*)\big) \cap \big(\nu^* \circ (\alpha^* \wedge \theta^*)\big) - \alpha^* .$$

This proves that item (4) of the theorem fails to hold.

Now we turn our attention to item (1). If $\mathbf{2} = \langle\{0,1\}; +\rangle$ is the 2-element join semilattice, $\theta, \nu \in \mathrm{Con}(\mathbf{2} \times \mathbf{2})$ are the projection kernels, and $\alpha \in \mathrm{Con}(\mathbf{2} \times \mathbf{2})$ is the kernel of the homomorphism $+\colon \mathbf{2}^2 \to \mathbf{2}$, then θ, ν and α generate a special $\mathbf{D}_2$. Therefore the variety of semilattices does not satisfy the idempotent Maltsev condition in (3). This shows that (3) $\Longrightarrow$ (1). To complete the proof, we show that (1) $\Longrightarrow$ (3).

Both (1) and (3) are statements about the idempotent Maltsev conditions satisfied by $\mathcal{V}$. The idempotent reduct $\mathrm{Id}(\mathcal{V})$ satisfies the same idempotent Maltsev conditions as $\mathcal{V}$, so to prove (1) $\Longrightarrow$ (3) we may (and do) assume that $\mathcal{V}$ is idempotent.

Suppose that (3) fails. Then $\mathcal{V}$ does not have terms satisfying the identities listed in (3), so the method that we used to construct those terms in our proof that (2) $\Longrightarrow$ (3) must fail. Recall from that part of the proof that we defined $\mathbf{F} = \mathbf{F}_{\mathcal{V}}(x, y, u, v)$, $\theta = \mathrm{Cg}^{\mathbf{F}}(x, u)$, $\mu = \mathrm{Cg}^{\mathbf{F}}(u, v)$, $\nu = \mathrm{Cg}^{\mathbf{F}}(x, y)$, and $\delta = \mathrm{Cg}^{\mathbf{F}}(y, v)$. We argued that if

$$(x, v) \in [(\theta \vee \delta) \wedge (\nu \vee \mu)] \vee \delta \vee \mu ,$$

then (3) holds. Therefore, in our current situation, the congruence $\alpha := [(\theta \vee \delta) \wedge (\nu \vee \mu)] \vee \delta \vee \mu$ does not contain the pair (x, v). Our plan is to use this fact to construct a nontrivial algebra in $\mathcal{V}$ that has a compatible semilattice operation.

Let $X = \{x, y, u, v\}$, so that $\mathbf{F} = \mathbf{F}_{\mathcal{V}}(X)$. Define $\mathbf{G} = \mathbf{F}_{\mathcal{V}}(x, v)$ to be the subalgebra of $\mathbf{F}$ generated by $\{x, v\}$. The functions $\tau_i\colon X \to X$ for $i = 1, 2$, defined by $\tau_1(x) = \tau_1(y) = x = \tau_2(x) = \tau_2(u)$ and $\tau_1(u) = \tau_1(v) = v = \tau_2(y) = \tau_2(v)$ are transversals for the equivalence relations $E_1 = \mathrm{Eq}^X\big((x, y), (u, v)\big)$ and $E_2 = \mathrm{Eq}^X\big((x, u), (y, v)\big)$ respectively. These transversals induce homomorphisms $h_i\colon \mathbf{F} \to \mathbf{F}$ whose kernels are $\theta \vee \delta$ and $\nu \vee \mu$ respectively (according to Lemma 3.6 (1)) and whose images are both $\mathbf{G}$. Therefore the kernel of

$$h := h_1 \times h_2\colon \mathbf{F} \to \mathbf{G} \times \mathbf{G}\colon t(x, y, u, v) \mapsto \big(t(x, x, v, v), t(x, v, x, v)\big)$$

is $[(\theta \vee \delta) \wedge (\nu \vee \mu)]$. The idempotence of $\mathcal{V}$ implies that this homomorphism is surjective. Namely, if $(p, q) = \big(p(x, v), q(x, v)\big) \in G \times G$, then $(p \circ q)(x, y, u, v) := p\big(q(x, y), q(u, v)\big) \in F$ is a preimage of (p, q) under h. This shows that $\mathbf{G} \times \mathbf{G}$ is generated by the four elements $h(x) = (x, x)$, $h(y) = (x, v)$, $h(u) = (v, x)$, and $h(v) = (v, v)$. Since the kernel of h is contained in $\alpha = [(\theta \vee \delta) \wedge (\nu \vee \mu)] \vee \delta \vee \mu$, we may use h to recast our earlier assumption that $(x, v) \notin \alpha$ as a statement about $\mathbf{G} \times \mathbf{G}$. That assumption is equivalent to the assumption that $h(x, v) = \big((x, x), (v, v)\big)$ does not belong to the congruence

$$\begin{aligned} h(\alpha) &= h\big([(\theta \vee \delta) \wedge (\nu \vee \mu)] \vee \delta \vee \mu\big) \\ &= \mathrm{Cg}^{\mathbf{G} \times \mathbf{G}}\big(h(\delta \vee \mu)\big) \\ &= \mathrm{Cg}^{\mathbf{G} \times \mathbf{G}}\big(h(y, v), h(u, v)\big) \\ &= \mathrm{Cg}^{\mathbf{G} \times \mathbf{G}}\big(((x, v), (v, v)), ((v, x), (v, v))\big) . \end{aligned}$$

Therefore, if we let $\beta = h(\alpha)$, then we have $\beta = \mathrm{Cg}^{\mathbf{G}\times\mathbf{G}}(Y \times Y)$ where $Y = \{(x,v),(v,x),(v,v)\}$, and our earlier assumption that $(x,v) \notin \alpha$ in $\mathbf{F}$ is equivalent to the assumption that $\big((x,x),(v,v)\big) \notin \beta$ in $\mathbf{G}\times\mathbf{G}$.

For any $(p,q) \in G^2$ there is an $r \in G$ such that $(p,q)\ \beta\ (r,r)$. Indeed, if $p = p(x,v)$ and $q = q(x,v)$, then for $r = p\big(q(x,v),v\big)$ we have

$$
\begin{aligned}
(p,q) &= \big(p(x,v), q(x,v)\big)\\
&= \big(p(q(x,x),q(v,v)),\ p(q(x,v),q(x,v))\big)\\
&= (p\circ q)\big((x,x),(x,v),(v,x),(v,v)\big)\\
&\equiv_\beta (p\circ q)\big((x,x),(v,v),(v,v),(v,v)\big)\\
&= \big(p(q(x,v),q(v,v)),\ p(q(x,v),q(v,v))\big)\\
&= \big(p(q(x,v),v),\ p(q(x,v),v)\big)\\
&= (r,r)\,.
\end{aligned}
\tag{5.8}
$$

Of course, the element r need not be unique, since it may happen that $(p,q)\ \beta\ (r,r)\ \beta\ (s,s)$ for some $s \neq r$. To account for this nonuniqueness, let γ be the congruence on $\mathbf{G}$ defined by

$$r\ \gamma\ s \iff (r,r)\ \beta\ (s,s)\,.$$

This is indeed a congruence, as it is $\iota^{-1}(\beta)$ for $\iota\colon \mathbf{G} \to \mathbf{G}^2\colon p \mapsto (p,p)$ equal to the diagonal embedding. Write $\overline{G}$ for G/γ and $\overline{p}$ for p/γ if $p \in G$. In this notation, we have shown in (5.8) that there is a well defined function

$$*\colon G\times G \to \overline{G}\colon (p,q) \mapsto \overline{r} \quad \text{if } (p,q)\ \beta\ (r,r)\,.$$

Explicitly, the value of $p*q$ is $\overline{p\big(q(x,v),v\big)}$, as we have shown.

If $q\ \gamma\ q'$, then $p\big(q(x,v),v\big)\ \gamma\ p\big(q'(x,v),v\big)$, since p is a term and γ is a congruence. Hence

$$p*q = \overline{p\big(q(x,v),v\big)} = \overline{p\big(q'(x,v),v\big)} = p*q'\,,$$

showing that left $*$-multiplication by any $p \in G$ is compatible with γ. Since the generators of β are invariant under switching coordinates, the property $(p,q)\ \beta\ (r,r)$ is equivalent to $(q,p)\ \beta\ (r,r)$, so $p*q = q*p$. This commutativity implies that right $*$-multiplication is also compatible with γ. Consequently $*$ induces a well defined binary operation (also called $*$) on $\overline{G}$:

$$*\colon \overline{G}\times\overline{G} \to \overline{G}\colon (\overline{p},\overline{q}) \mapsto \overline{r} \quad \text{if } (p,q)\ \beta\ (r,r)\,.$$

We have already established the commutativity of this operation. The idempotence of $*$ follows from the fact that $(p,p)\ \beta\ (p,p)$ for any $p \in G$. Associativity is clear, too, since

$$(\overline{p}*\overline{q})*\overline{r} = \overline{p\big(q(r(x,v),v),v\big)} = \overline{p}*(\overline{q}*\overline{r})\,.$$

Thus $*$ is a semilattice operation on the set $\overline{G}$. Moreover, $*$ is a compatible semilattice operation of the algebra $\overline{\mathbf{G}} := \mathbf{G}/\gamma$. For suppose that f is an n-ary basic operation of $\overline{G}$, and that $\overline{p}_i, \overline{q}_i \in \overline{G}$ are chosen arbitrarily. Choose r_i such that $(p_i,q_i)\ \beta\ (r_i,r_i)$ for all i. Then, since β is a congruence, $\big(f(\mathbf{p}),f(\mathbf{q})\big)\ \beta\ \big(f(\mathbf{r}),f(\mathbf{r})\big)$. This shows that

$$f(\overline{\mathbf{p}})*f(\overline{\mathbf{q}}) = f(\overline{\mathbf{r}}) = f(\overline{\mathbf{p}}*\overline{\mathbf{q}})\,.$$

Finally (and critically), $\overline{\mathbf{G}}$ is nontrivial. Indeed, as $\big((x,x),(v,v)\big) \notin \beta$, we have that $(x,v) \notin \gamma$, so $\overline{x}$ and $\overline{v}$ are distinct elements of $\overline{\mathbf{G}}$.

This proves that $\overline{\mathbf{G}} \in \mathcal{V}$ is a nontrivial algebra with a compatible semilattice operation. Theorem 5.6 applies to show that the total relation on $\overline{\mathbf{G}}$ is rectangular. From Theorem 5.25 we derive that any idempotent Maltsev condition satisfied by

$\mathcal{V}$ is also satisfied by the variety of semilattices. Thus (1) fails, and the proof is complete. □

In Theorem 8.11 we will learn that if $\mathcal{V}$ satisfies an idempotent Maltsev condition that fails in the variety of semilattices, then $\mathcal{V}$ omits all $\mathbf{D}_2$'s, not just the special ones.

CHAPTER 6

A Theory of Solvability

A normal subgroup N of a group G is **solvable** if there is a finite chain of normal subgroups of G,

$$\{1\} = N_0 \subseteq N_1 \subseteq \cdots \subseteq N_k = N\,,$$

such that each factor N_{i+1}/N_i is abelian, or equivalently such that $[N_{i+1}, N_{i+1}] \subseteq N_i$ for all i. An interval $I[M, N]$ in the normal subgroup lattice of G is a **solvable interval** if N/M is a solvable normal subgroup of G/M. Normal subgroups M and N of G are **solvably related**, written $M \overset{s}{\sim} N$, if they belong to a solvable interval. We will call the relation $\overset{s}{\sim}$ "the solvability relation", and refer to the study of this solvability relation as the "theory of solvability" for groups. Among the important features of the theory of solvability for groups are:

- The solvability relation is a congruence on the normal subgroup lattice of any group.
- The solvability relation is preserved by homomorphisms in the sense that if M, N and P are normal subgroups of G with $P \leq M \cap N$, then M is solvably related to N if and only if M/P is solvably related to N/P in G/P.
- If N is a minimal normal subgroup of G, then N is solvable if and only if it is abelian.

The ring-theoretic concept that corresponds directly to that of a solvable normal subgroup is that of a nilpotent ideal. In the early days of ring theory the concept of solvability was developed in analogy with the group theoretical concept via the Wedderburn radical (= the largest nilpotent ideal, if it exists). Other radicals soon made their appearance, which permitted new approaches to solvability not bound to *finite* chains of congruences with abelian factors. Moreover, in the case of some radicals, the resulting "solvability theory" was no longer bound to the abelian property. For example, if $\mathbf{R}$ is a subdirectly irreducible ring that is not simple, then the Brown–McCoy radical (= the intersection of maximal two-sided ideals) contains the monolith *whether or not the monolith is abelian.* Thus, it can happen that a minimal nonabelian congruence is contained in the Brown–McCoy radical, hence is "solvable" in the Brown–McCoy sense.

A third approach to solvability, valid for arbitrary finite algebras, is developed by D. Hobby and R. McKenzie in Chapter 7 of [**34**]. This approach does not use chains of congruences with abelian factors nor does it use radicals to define the solvability concept. Rather, one starts by defining a pair $(0,1) \in A^2$ from an algebra $\mathbf{A}$ to be a **2-snag** if there is a polynomial $p(x,y) \in \mathrm{Pol}_2(\mathbf{A})$ such that $p(0,0) = p(0,1) = p(1,0) = 0$ and $p(1,1) = 1$. Hobby and McKenzie prove that if $\mathbf{A}$ is a finite algebra and $\alpha \prec \beta$ are congruences on $\mathbf{A}$, then $\mathbf{C}(\beta, \beta; \alpha)$ holds if and only if $\beta - \alpha$ contains no 2-snag. From this it is clear that a chain of congruences:

$$\alpha = \alpha_0 \prec \alpha_1 \prec \cdots \prec \alpha_k = \beta$$

has abelian factors α_{i+1}/α_i if and only if $\beta - \alpha$ contains no 2-snag. Starting from this observation, it can be argued that the relation $\alpha \overset{s}{\sim} \beta$, defined by the condition that α and β contain the same 2-snags, is a complete congruence on $\mathbf{Con}(\mathbf{A})$ for any locally finite algebra.

Our aim in this chapter is to synthesize ideas from the solvability theories for groups, rings, and finite algebras in order to develop a solvability theory for a wide class of varieties. The theory resembles the theory for groups in that it is defined in terms of chains of congruences with abelian factors. However, as we will show in Section 6.3, the theory may viewed as an extension of the theory of solvability for rings that is based on the prime radical (= the intersection of all prime congruences), since for rings our theory defines an ideal to be solvable precisely when it is contained in the prime radical. Finally, anyone familiar with the theory developed by Hobby and McKenzie in [**34**] will recognize that our "solvability obstructions" (Definition 6.5) play the role of their 2-snags, while Theorems 6.17, 6.23 and 6.25 of this chapter are analogues of Theorem 7.9 (3), Corollary 7.13 and Lemma 6.5 of [**34**].

6.1. Varieties with a Weak Difference Term

There is an obstacle to the development of a very general solvability theory that we must deal with first. Any absolutely free algebra in any given signature is abelian in the sense of Chapter 2. If there were a solvability theory that applied to all such algebras, where the solvability relation is a congruence preserved by homomorphisms, then such a theory must identify every congruence on every algebra as solvable. Such a theory would, of course, be trivial. We cannot avoid this type of problem even if we restrict our attention to varieties satisfying a nontrivial idempotent Maltsev condition, since there exist algebras (such as $\mathbf{A} = \langle [0,1]; \frac{x+y}{2} \rangle$, the unit interval of the real line under the averaging operation) that are abelian in the sense of Chapter 2, have a Taylor term ($f(x,y) = \frac{x+y}{2}$), but which have simple nonabelian homomorphic images (in this case, $\mathbf{A}/\theta$ where $\theta = [0,1) \times [0,1) \cup \{(1,1)\}$). This phenomenon, that solvable algebras have nonsolvable homomorphic images, prevents the development of a powerful solvability theory.

As it happens, there is a weakest idempotent Maltsev condition that guarantees that homomorphic images of solvable algebras are solvable, and this Maltsev condition is a strong enough hypothesis for the theorems that we have in mind. We start this chapter by discussing the idempotent Maltsev condition on which the solvability theory is based, and then we discuss the theory.

DEFINITION 6.1. Let $\mathcal{V}$ be a variety. A ternary term $d(x,y,z)$ is a **weak difference term** for $\mathcal{V}$ if whenever $\mathbf{A} \in \mathcal{V}$, $\theta \in \mathrm{Con}(\mathbf{A})$ and $(a,b) \in \theta$ it is the case that

$$d(a,a,b) \equiv b \equiv d(b,a,a) \pmod{[\theta,\theta]}.$$

The following theorem is included to show that our development of a solvability theory only for varieties with a weak difference term is reasonable.

THEOREM 6.2. *The class of varieties with a weak difference term is definable by a nontrivial idempotent Maltsev condition. This idempotent Maltsev condition is the weakest conjunction of idempotent Maltsev conditions that implies that homomorphic images of solvable algebras are solvable. It is also the weakest conjunction of idempotent Maltsev conditions that implies that abelian algebras are affine.*

PROOF. The first and third statements are proved in Theorem 4.8 of [**53**]. We prove the second only.

If $\mathcal{V}$ has a weak difference term, then it can be shown (by using Lemma 6.8 and following the same argument that works for groups) that homomorphic images of solvable algebras are solvable.

Conversely assume that $\mathcal{V}$ satisfies some conjunction of idempotent Maltsev conditions that implies that homomorphic images of solvable algebras are solvable. At least one of these Maltsev conditions must be nontrivial, since there exist solvable algebras with nonsolvable homomorphic images (e.g. $\langle [0,1]; \frac{x+y}{2}\rangle$). The idempotent reduct $\mathrm{Id}(\mathcal{V})$ of $\mathcal{V}$ satisfies the same idempotent Maltsev conditions as $\mathcal{V}$, so homomorphic images of solvable algebras in $\mathrm{Id}(\mathcal{V})$ are again solvable. Let $\mathcal{A}$ be the subvariety of $\mathrm{Id}(\mathcal{V})$ that is generated by the abelian algebras in $\mathrm{Id}(\mathcal{V})$. Since the class of abelian algebras in any language is axiomatizable by the universal Horn sentences of the term condition, this class is closed under the formation of subalgebras and products. From our assumption that homomorphic images of solvable algebras are solvable we get that $\mathcal{A}$ consists of solvable algebras. In particular, $\mathcal{A}$ contains no subvariety term equivalent to the variety of semilattices. Since $\mathcal{A}$ also satisfies the nontrivial idempotent Maltsev conditions satisfied by $\mathcal{V}$, it follows from Lemma 2.5 of [**47**] that $\mathcal{A}$ must satisfy an idempotent Maltsev condition that fails in the variety of semilattices. From Theorem 4.10 of [**53**], it follows that the variety $\mathcal{A}$ is affine, and in particular congruence permutable. If $d(x, y, z)$ is a Maltsev operation for $\mathcal{A}$, then it will be a weak difference term for $\mathcal{V}$, as is explained in the proof of Theorem 4.8 of [**53**]. □

This theorem has a corollary which will prove useful in what follows.

COROLLARY 6.3. *If a variety has a Hobby–McKenzie term, then it has a weak difference term.*

PROOF. Theorem 4.10 of [**53**] proves that if $\mathcal{V}$ satisfies an idempotent Maltsev condition that fails in the variety of semilattices, then abelian algebras in $\mathcal{V}$ are affine. Theorem 5.25 proves that the hypothesis of this statement is equivalent to the existence of a Hobby–McKenzie term, while Theorem 6.2 proves that the conclusion of the statement implies the existence of a weak difference term. □

The concept of a weak difference term entered mathematics during the development of commutator theory. Recall from the introduction that A. I. Maltsev showed in [**63**] that a variety $\mathcal{V}$ has permuting congruences if and only if it has a term $d(x, y, z)$ such that $\mathcal{V} \models d(x, x, y) \approx y \approx d(y, x, x)$. In [**31**], C. Herrmann proved that abelian algebras in a congruence modular variety $\mathcal{V}$ are affine by constructing a term $d(x, y, z)$ from the Day terms for congruence modularity and showing that $d(x, y, z)$ is a Maltsev term for any abelian algebra in $\mathcal{V}$.[1] Then

[1]This fact, together with the fact that the class of congruence modular varieties is definable by an idempotent Maltsev condition, already implies that $d(x, y, z)$ is a weak difference term for $\mathcal{V}$, as the proof of Theorem 6.2 shows.

H. P. Gumm observed in [**27**] that the term constructed by Herrmann has the stronger properties that the identity $d(x,x,y) \approx y$ holds in $\mathcal{V}$, and $d(b,a,a) \equiv b \pmod{[\theta,\theta]}$ for any $\mathbf{A} \in \mathcal{V}$, $\theta \in \mathrm{Con}(\mathbf{A})$, and $(a,b) \in \theta$. Such a term is now called a **difference term**. Difference terms play a central role in most developments of modular commutator theory and in some extensions of the theory to nonmodular varieties (see [**45, 57, 59**]). Weak difference terms were first identified in [**34**]. They occur far more frequently than ordinary difference terms, and are equally useful in most circumstances.

EXAMPLES 6.4. As noted in Theorem 6.2, any variety with a weak difference term satisfies a nontrivial idempotent Maltsev condition, and therefore has a Taylor term. The converse is not true in general, as we explained in the first paragraph of this section, but the converse does hold for certain classes of varieties. We describe some of these classes now, thereby providing a number of examples of varieties with weak difference terms.

Theorem 9.6 of [**34**] proves the converse for locally finite varieties: that is, any locally finite variety that has a Taylor term has a weak difference term.

Any idempotent abelian (or solvable) variety that has a Taylor term has a weak difference term. (The proof of this is contained in the proof of Theorem 6.2.)

A variety of semigroups that has a Taylor term has a weak difference term. Indeed, suppose that $f(x_1,\ldots,x_n) = x_{i_0}\cdots x_{i_k}$ is a Taylor term for $\mathcal{V}$. Then necessarily $k \geq 1$. The idempotence of f implies that $\mathcal{V} \models x^{k+1} \approx x$. Any semigroup $\mathbf{S}$ satisfying this equation has the property that $(s^k)^2 = s^{k-1}s^{k+1} = s^{k-1}s = s^k$ for any $s \in S$, so k-th powers of elements of $\mathbf{S}$ are idempotent. If $e, f \in S$ are idempotent, and $ef \neq fe$, then it can be shown that the subsemigroup of $\mathbf{S}$ generated by $\{e,f\}$ has a homomorphism onto a nontrivial left zero or right zero semigroup. But a nontrivial left zero semigroup fails the i_0-th Taylor identity for $f(x_1,\ldots,x_n) = x_{i_0}\cdots x_{i_k}$, and a nontrivial right zero semigroup fails the i_k-th Taylor identity. It follows that $\mathcal{V}$ has no nontrivial left or right zero semigroups, and hence that no $\mathbf{S} \in \mathcal{V}$ has a pair of noncommuting idempotents. This fact may be expressed by saying that the identity $x^ky^k \approx y^kx^k$ holds in $\mathcal{V}$. Now we argue that the term $d(x,y,z) = xy^{2k-1}z$ is a weak difference term for the variety of semigroups axiomatized by $x^{k+1} \approx x$ and $x^ky^k \approx y^kx^k$. To justify this, it suffices to show that if $\mathbf{A}$ is a semigroup in $\mathcal{V}$ and $a, b \in A$ are elements for which $\theta := \mathrm{Cg}^{\mathbf{A}}(a,b)$ is abelian, then $d(a,a,b) = b = d(b,a,a)$. To show that $b = d(a,a,b) = a^{2k}b = a^kb$ and $b = ba^k$ it is enough to prove that $a^k = b^k$. Since $e = a^k$ and $f = b^k$ are idempotent and $(e,f) = (a^k,b^k) \in \theta$, it suffices to prove that idempotents related by an abelian congruence are equal. The θ,θ-term condition applied to $ee\underline{f} = ef = ef\underline{f}$ allows us to replace the underlined f with e and obtain $eee = efe$. Since idempotents commute, this is equivalent to $e = ef$. The same argument shows that $f = ef$, so $e = f$.

It is easily seen that any congruence n-permutable variety satisfies a nontrivial idempotent Maltsev condition, hence has a Taylor term. Less obvious is the theorem of P. Lipparini in [**58**] that any congruence n-permutable variety has a weak difference term. (This fact can be derived from Corollary 6.3.)

6.2. ∞-Solvability

We now begin the development of a solvability theory for varieties with a weak difference term. Any successor ordinal, considered as a lattice under its usual ordering, is a meet continuous lattice. By a **continuous sequence of congruences** on an algebra $\mathbf{A}$ we mean a meet continuous lattice homomorphism $\sigma : \kappa + 1 \to \mathbf{Con}(\mathbf{A})$. We write such a sequence as $(\sigma_\lambda)_{\lambda<\kappa+1}$ where $\sigma_\lambda := \sigma(\lambda)$. It is not hard to verify that $(\sigma_\lambda)_{\lambda<\kappa+1}$ is a continuous sequence of congruences if and only if $\sigma_\mu \leq \sigma_\lambda$ for $\mu < \lambda$ and $\sigma_\lambda = \bigcup_{\mu<\lambda} \sigma_\mu$ for all limit ordinals λ.

DEFINITION 6.5. Let $\mathbf{A}$ be an algebra with congruences α, β, θ, θ_λ, δ and δ'.

If $\delta \leq \theta$ and $\mathbf{C}(\theta, \theta; \delta)$ holds then we say that **θ is abelian over δ**, and write $\delta \lhd \theta$.

If κ is an ordinal, and $(\theta_\lambda)_{\lambda<\kappa+1}$ is a continuous sequence of congruences such that $\theta_\lambda \lhd \theta_{\lambda+1}$ for all $\lambda < \kappa$, then we say that **θ_κ is κ-step solvable over θ_0**.

We say that **θ is ∞-solvable over δ** if θ is κ-step solvable over δ for some κ. We may denote this by $\delta \lll \theta$. If $0 \lll \theta$, then we say that **θ is ∞-solvable**. If $0 \lll 1$, then we say that **$\mathbf{A}$ is ∞-solvable**.

If $\alpha \wedge \beta \lll \alpha \vee \beta$, then we say that **$\alpha$ is solvably related to β** and denote this by $\alpha \overset{s}{\sim} \beta$.

If $\delta < \theta$, and $\delta \lhd \delta' \leq \theta$ implies $\delta = \delta'$, then we call the congruence interval $I[\delta, \theta]$ a **solvability obstruction** and we denote this by $\delta \blacktriangleleft \theta$.

Note that for comparable congruences $\delta \leq \theta$ the relations $\delta \lll \theta$ and $\delta \overset{s}{\sim} \theta$ are the same.

EXAMPLE 6.6. If $\mathcal{V}$ is a variety whose members have the property that their principal congruences are abelian, then every member of $\mathcal{V}$ is ∞-solvable. For if $\mathbf{A} \in \mathcal{V}$, $(a_\lambda)_{\lambda<\kappa}$ is an enumeration of $\mathbf{A}$, and θ_λ is the congruence generated by $\{(a_0, a_\mu) \mid \mu < \lambda\}$, then it is not hard to see that

(i) $\theta_0 = 0$ and $\theta_\kappa = 1$,
(ii) $\theta_\lambda \lhd \theta_{\lambda+1}$ for all $\lambda < \kappa$, and
(iii) $\theta_\lambda = \bigcup_{\mu<\lambda} \theta_\mu$ for all limit ordinals.

Items (i) and (iii) follow immediately from the definition of θ_λ. For (ii), note that $\theta_{\lambda+1} = \mathrm{Cg}^{\mathbf{A}}(a_0, a_\lambda) \vee \theta_\lambda$. Thus $\theta_{\lambda+1}/\theta_\lambda$ is a principal (hence abelian) congruence on $\mathbf{A}/\theta_\lambda$. This implies that $\theta_\lambda \lhd \theta_{\lambda+1}$ for all $\lambda < \kappa$.

An example of a variety in which principal congruences are abelian is the variety $\mathcal{V}$ of nonunital rings satisfying $x^2 \approx 0$. This identity implies that $0 \approx (x+y)^2 \approx x^2+xy+yx+y^2 \approx xy+yx$, so the anticommutative law $xy \approx -yx$ holds in $\mathcal{V}$. This law implies that, for any $\mathbf{R} \in \mathcal{V}$, all principal ideals have the form $(a) = R\cdot a$. Hence $(a)^2 = RaRa = \sum\{ra \cdot sa = -rsa^2 \mid r, s \in R\} = (0)$. Since congruences correspond to ideals, and the commutator operation corresponds to ideal multiplication, this is indeed a variety in which principal congruences are abelian. Hence $\mathcal{V}$ consists entirely of ∞-solvable algebras.

However, there is no cardinal bound on solvability degree in $\mathcal{V}$. In fact, if κ is any infinite cardinal, then $\mathcal{V}$ contains rings of solvability degree exactly κ. One such ring, $\mathbf{R}_\kappa$, may be presented by generators and relations relative to $\mathcal{V}$ as follows. Let S be an infinite set of cardinality κ. Take as generators for $\mathbf{R}_\kappa$ all symbols x_U where U is a nonempty subset of S. Impose the relations

- $2x_U = 0$ for all nonempty $U \subseteq S$,
- $x_U \cdot x_V = 0$ if $U \cap V \neq \emptyset$, and
- $x_U \cdot x_V = x_{U\cup V}$ if $U \cap V = \emptyset$.

These relations force each element of $\mathbf{R}_\kappa$ to be expressible as a finite sum of generators $x_{U_1} + \cdots + x_{U_n}$ where $U_i \neq U_j$ if $i \neq j$. We leave it as an exercise to prove that the ring $\mathbf{R}_\kappa$ is κ-step solvable, but not λ-step solvable for any ordinal $\lambda < \kappa$.

LEMMA 6.7. *Let $\mathcal{V}$ be a variety with a weak difference term. If $\mathbf{A} \in \mathcal{V}$ has a tolerance T and a congruence δ such that $\mathbf{C}(T, T; \delta)$ holds, then $T' := \delta \circ T \circ \delta$ is the smallest congruence on $\mathbf{A}$ containing $T \cup \delta$. Furthermore, $\mathbf{C}(T', T'; \delta)$ holds.*

PROOF. We know from the basic properties of the centralizer relation (see Theorem 2.19) that if $\mathbf{C}(T, T; \delta)$ holds and $T' := \delta \circ T \circ \delta$, then $\mathbf{C}(T', T'; \delta)$ holds. Since $(T \cup \delta) \subseteq T' = \delta \circ T \circ \delta \subseteq \mathrm{Cg}^{\mathbf{A}}(T \cup \delta)$, the only thing left to show is that T' is a congruence. For this we only need to establish that $T' \circ T' \subseteq T'$.

Let $\theta = \mathrm{Cg}^{\mathbf{A}}(T')$. Since $\mathbf{C}(T', T'; \delta)$ holds, it follows from Theorems 3.24 and 6.2 that $\mathbf{C}(\theta, \theta; \delta)$ holds. If $(a, c) \in T' \circ T'$, then there is an element b such that $(a, b), (b, c) \in T' \subseteq \theta$. Thus

$$a \equiv_{[\theta,\theta]} d(a, b, b) \equiv_{T'} d(b, b, c) \equiv_{[\theta,\theta]} c$$

and $[\theta, \theta] \leq \delta$ imply that $(a, c) \in \delta \circ T' \circ \delta = T'$. Since $(a, c) \in T' \circ T'$ was chosen arbitrarily, $T' \circ T' \subseteq T'$. □

LEMMA 6.8. *Let $\mathcal{V}$ be a variety with a weak difference term. If $\mathbf{A} \in \mathcal{V}$ has congruences θ and δ such that $[\theta, \theta] \leq \delta$, then $\mathbf{C}(\theta, \theta; \delta)$ holds.*

PROOF. Assume that $[\theta, \theta] \leq \delta$. If $\mathbf{C}(\theta, \theta; \delta)$ does not hold, then there are a polynomial $p(\mathbf{x}, \mathbf{y})$ and tuples $\mathbf{a}\ \theta\ \mathbf{b}$ and $\mathbf{u}\ \theta\ \mathbf{v}$ such that

$$p(\mathbf{a}, \mathbf{u}) \equiv_\delta p(\mathbf{a}, \mathbf{v})$$

and

$$p(\mathbf{b}, \mathbf{u}) \not\equiv_\delta p(\mathbf{b}, \mathbf{v})\,. \tag{6.1}$$

Let $p'(\mathbf{x}, \mathbf{y}) = d\big(p(\mathbf{x}, \mathbf{y}), p(\mathbf{x}, \mathbf{u}), p(\mathbf{b}, \mathbf{u})\big)$. Using that d is an idempotent term that is Maltsev on θ-classes modulo $[\theta, \theta]$, we get that

$$p'(\mathbf{a}, \underline{\mathbf{u}}) = d\big(p(\mathbf{a}, \mathbf{u}), p(\mathbf{a}, \mathbf{u}), p(\mathbf{b}, \mathbf{u})\big) \equiv_{[\theta,\theta]} p(\mathbf{b}, \mathbf{u}) = p'(\mathbf{b}, \underline{\mathbf{u}})\,.$$

Since $\mathbf{C}(\theta, \theta; [\theta, \theta])$ holds, we can change the underlined $\mathbf{u}$ to $\mathbf{v}$ without changing the fact that the left hand side is $[\theta, \theta]$-related to the right hand side:

$$p'(\mathbf{a}, \underline{\mathbf{v}}) \equiv_{[\theta,\theta]} p'(\mathbf{b}, \underline{\mathbf{v}})\,. \tag{6.2}$$

Moreover we have that

$$\begin{aligned} p'(\mathbf{a}, \mathbf{v}) &= d\big(p(\mathbf{a}, \mathbf{v}), p(\mathbf{a}, \mathbf{u}), p(\mathbf{b}, \mathbf{u})\big) \\ &\equiv_\delta d\big(p(\mathbf{a}, \mathbf{u}), p(\mathbf{a}, \mathbf{u}), p(\mathbf{b}, \mathbf{u})\big) \equiv_{[\theta,\theta]} p(\mathbf{b}, \mathbf{u})\,. \end{aligned} \tag{6.3}$$

Here

$$d\big(p(\mathbf{a}, \mathbf{v}), p(\mathbf{a}, \mathbf{u}), p(\mathbf{b}, \mathbf{u})\big) \equiv_\delta d\big(p(\mathbf{a}, \mathbf{u}), p(\mathbf{a}, \mathbf{u}), p(\mathbf{b}, \mathbf{u})\big)$$

is a consequence of the fact that $p(\mathbf{a}, \mathbf{v}) \equiv_\delta p(\mathbf{a}, \mathbf{u})$, and

$$d\big(p(\mathbf{a}, \mathbf{u}), p(\mathbf{a}, \mathbf{u}), p(\mathbf{b}, \mathbf{u})\big) \equiv_{[\theta,\theta]} p(\mathbf{b}, \mathbf{u})$$

is a consequence of the facts that $\mathbf{a}\ \theta\ \mathbf{b}$ and that d is Maltsev on θ-classes modulo $[\theta, \theta]$. We also have that

$$p'(\mathbf{b}, \mathbf{v}) = d\big(p(\mathbf{b}, \mathbf{v}), p(\mathbf{b}, \mathbf{u}), p(\mathbf{b}, \mathbf{u})\big) \equiv_{[\theta,\theta]} p(\mathbf{b}, \mathbf{v})\,, \tag{6.4}$$

since $\mathbf{u}\ \theta\ \mathbf{v}$ and d is Maltsev on θ-classes modulo $[\theta,\theta]$. Combining lines (6.2), (6.3), and (6.4) yields

$$p(\mathbf{b},\mathbf{u}) \equiv_\delta p'(\mathbf{a},\mathbf{v}) \equiv_{[\theta,\theta]} p'(\mathbf{b},\mathbf{v}) \equiv_{[\theta,\theta]} p(\mathbf{b},\mathbf{v})\,,$$

or simply $p(\mathbf{b},\mathbf{u}) \equiv_\delta p(\mathbf{b},\mathbf{v})$ (since $[\theta,\theta] \leq \delta$). This contradicts line (6.1), thereby completing the proof. □

The following corollary will be used in Section 9.2.

COROLLARY 6.9. *Let $\mathcal{V}$ be a variety with a weak difference term. Assume that T is an abelian tolerance on some $\mathbf{A} \in \mathcal{V}$, and that a and b are distinct T-related elements. If ρ is a congruence on $\mathbf{A}$ that is maximal with respect to $(a,b) \notin \rho$, then T/ρ is a nontrivial abelian congruence on the subdirectly irreducible algebra $\mathbf{A}/\rho$.*

PROOF. It follows from Lemma 6.7 (using $\delta = 0$) that T is an abelian congruence on $\mathbf{A}$. Then, using Lemma 6.8 in the case where $\theta = T$ and $\delta = \rho$, we obtain that $\mathbf{C}(T,T;\rho)$ holds, so T/ρ is a nontrivial abelian tolerance on the subdirectly irreducible algebra $\mathbf{A}/\rho$. Using Lemma 6.7, we find that T/ρ is a congruence. □

LEMMA 6.10. *Let $\mathcal{V}$ be a variety with a weak difference term. Assume that $\mathbf{A} \in \mathcal{V}$ has congruences α, β, and γ.*

(1) *If $\alpha \lhd \beta$, then $(\alpha \wedge \gamma) \lhd (\beta \wedge \gamma)$ and $(\alpha \vee \gamma) \lhd (\beta \vee \gamma)$.*
(2) *If $\alpha \lll \beta$, then $(\alpha \wedge \gamma) \lll (\beta \wedge \gamma)$ and $(\alpha \vee \gamma) \lll (\beta \vee \gamma)$.*

PROOF. To prove the part of item (1) that concerns the compatibility of $\lhd$ with meet, recall that $\alpha \lhd \beta$ means that $\mathbf{C}(\beta,\beta;\alpha)$ holds. By item (1) of Theorem 2.19 we have that $\mathbf{C}(\beta\wedge\gamma,\beta\wedge\gamma;\alpha)$ holds. By item (8) of Theorem 2.19 we have that $\mathbf{C}(\beta\wedge\gamma,\beta\wedge\gamma;\gamma)$ holds. Item (6) of Theorem 2.19 now yields that $\mathbf{C}(\beta\wedge\gamma,\beta\wedge\gamma;\alpha\wedge\gamma)$ holds, which means that $(\alpha\wedge\gamma) \lhd (\beta\wedge\gamma)$.

To prove the compatibility of $\lhd$ with join, assume that $\mathbf{C}(\beta,\beta;\alpha)$ holds. Then we have $[\beta,\beta] \leq \alpha \leq \alpha\vee\gamma$, so by Lemma 6.8 we get that $\mathbf{C}(\beta,\beta;\alpha\vee\gamma)$ holds. Using Lemma 6.7 with $T = \beta$ and $\delta = \alpha\vee\gamma$ we deduce that $(\alpha\vee\gamma)\circ\beta\circ(\alpha\vee\gamma) = \beta\vee(\alpha\vee\gamma) = \beta\vee\gamma$ and that $\mathbf{C}(\beta\vee\gamma,\beta\vee\gamma;\alpha\vee\gamma)$ holds. This shows that $(\alpha\vee\gamma) \lhd (\beta\vee\gamma)$, so we proved item (1).

For (2), assume that $(\theta_\lambda)_{\lambda<\kappa+1}$ is a continuous sequence of congruences for which $\alpha = \theta_0$, $\beta = \theta_\kappa$ and $\theta_\lambda \lhd \theta_{\lambda+1}$ for all λ. The sequences $(\theta_\lambda \wedge \gamma)_{\lambda<\kappa+1}$ and $(\theta_\lambda \vee \gamma)_{\lambda<\kappa+1}$ are also continuous, since meeting or joining with the fixed element γ preserves order and unions of chains, so item (1) of this lemma implies that $(\theta_\lambda \wedge \gamma)_{\lambda<\kappa+1}$ and $(\theta_\lambda \vee \gamma)_{\lambda<\kappa+1}$ are sequences witnessing $(\alpha\wedge\gamma) \lll (\beta\wedge\gamma)$ and $(\alpha\vee\gamma) \lll (\beta\vee\gamma)$, respectively. □

THEOREM 6.11. *Let $\mathcal{V}$ be a variety with a weak difference term. If $\mathbf{A} \in \mathcal{V}$, then $\overset{s}{\sim}$ is a congruence on $\mathbf{Con}(\mathbf{A})$ that is compatible with the complete join operation.*

PROOF. Recall that $\alpha \overset{s}{\sim} \beta$ holds if and only if $(\alpha\wedge\beta) \lll (\alpha\vee\beta)$. It is easy to see that this relation is reflexive and symmetric. To show that it is transitive, assume that $\alpha \overset{s}{\sim} \beta \overset{s}{\sim} \gamma$. Then $(\alpha\wedge\beta) \lll (\alpha\vee\beta)$ and $(\beta\wedge\gamma) \lll (\beta\vee\gamma)$, so by the second statement of Lemma 6.10 we get that

$$(\alpha\wedge\beta\wedge\gamma) = (\alpha\wedge\beta)\wedge(\beta\wedge\gamma) \lll (\alpha\wedge\beta)\wedge(\beta\vee\gamma) = (\alpha\wedge\beta)$$

and

$$(\alpha\vee\beta) = (\alpha\vee\beta)\vee(\beta\wedge\gamma) \lll (\alpha\vee\beta)\vee(\beta\vee\gamma) = (\alpha\vee\beta\vee\gamma)\,.$$

Thus we have $(\alpha \wedge \beta \wedge \gamma) \lll (\alpha \wedge \beta) \lll (\alpha \vee \beta) \lll (\alpha \vee \beta \vee \gamma)$. The well-ordered chains of congruences that witness these three instances of ∞-solvability may be concatenated to produce a single such chain from $\alpha \wedge \beta \wedge \gamma$ to $\alpha \vee \beta \vee \gamma$, thus proving that $(\alpha \wedge \beta \wedge \gamma) \lll (\alpha \vee \beta \vee \gamma)$. Since $(\alpha \wedge \beta \wedge \gamma) \leq \alpha \wedge \gamma \leq \alpha \vee \gamma \leq (\alpha \vee \beta \vee \gamma)$, we can deduce that $(\alpha \wedge \gamma) \lll (\alpha \vee \gamma)$ if we show first that $\delta \leq \delta' \leq \theta' \leq \theta$ and $\delta \lll \theta$ imply $\delta' \lll \theta'$. We show this by using Lemma 6.10 again: joining $\delta \lll \theta$ with δ' we obtain

$$\delta' = (\delta' \vee \delta) \lll (\delta' \vee \theta) = \theta\,.$$

Meeting with θ' we find that

$$\delta' = (\delta' \wedge \theta') \lll (\theta \wedge \theta') = \theta'\,.$$

This reasoning completes the argument that $\stackrel{s}{\sim}$ is transitive, hence an equivalence relation, and also shows that subintervals of ∞-solvable intervals are again ∞-solvable. Hence $\stackrel{s}{\sim}$-classes are convex sublattices of $\mathbf{Con}(\mathbf{A})$. This implies that $\alpha \stackrel{s}{\sim} \beta$ if and only if α and β belong to some ∞-solvable interval. This reformulation of the definition of the $\stackrel{s}{\sim}$-relation makes it easy to see that it is a congruence: $\alpha \stackrel{s}{\sim} \beta$ if and only if $\delta := \alpha \wedge \beta \lll \alpha \vee \beta =: \theta$. But $\delta \lll \theta$ implies $\delta \wedge \gamma \lll \theta \wedge \gamma$ and $\delta \vee \gamma \lll \theta \vee \gamma$ by Lemma 6.10. Since $\alpha \wedge \gamma, \beta \wedge \gamma \in I[\delta \wedge \gamma, \theta \wedge \gamma]$ and $\alpha \vee \gamma, \beta \vee \gamma \in I[\delta \vee \gamma, \theta \vee \gamma]$, we conclude that $\alpha \stackrel{s}{\sim} \beta$ implies that $(\alpha \wedge \gamma) \stackrel{s}{\sim} (\beta \wedge \gamma)$ and $(\alpha \vee \gamma) \stackrel{s}{\sim} (\beta \vee \gamma)$.

It remains to show that $\stackrel{s}{\sim}$ is compatible with the complete join operation of $\mathbf{Con}(\mathbf{A})$. Suppose that $\alpha_i \stackrel{s}{\sim} \beta_i$ for all $i \in I$. Let $\delta_i = \alpha_i \wedge \beta_i$ and $\theta_i = \alpha_i \vee \beta_i$. Then $\delta_i \lll \theta_i$ for all i. If $\Delta := \bigvee_{i\in I} \delta_i$ and $\Theta := \bigvee_{i\in I} \theta_i$, then both $\bigvee_{i\in I} \alpha_i$ and $\bigvee_{i\in I} \beta_i$ belong to the interval $I[\Delta, \Theta]$. We will be done if we show that Θ is ∞-solvable over Δ.

Consider the collection $\mathcal{P}$ of all sequences of the form $\sigma = (\sigma_\lambda)_{\lambda<\kappa+1}$ where $\sigma_\lambda \in \mathrm{Con}(\mathbf{A})$ and

(0) $\sigma_0 = \Delta$, $\sigma_\kappa \in I[\Delta, \Theta]$,
(1) $\sigma_\lambda \lhd \sigma_{\lambda+1}$ for all $\lambda < \kappa$, and
(2) $\sigma_\lambda = \bigcup_{\mu<\lambda} \sigma_\mu$ for limit ordinals λ.

The elements of $\mathcal{P}$ are the continuous sequences of congruences that witness the ∞-solvability over Δ of some elements $\sigma_\kappa \in I[\Delta, \Theta]$. The set $\mathcal{P}$ is nonempty, since it contains the one-term sequence (Δ). $\mathcal{P}$ can be partially ordered by the rule: $\sigma \leq \sigma'$ if σ' is an extension of σ. That is, $\sigma \leq \sigma'$ if the domain of σ is contained in the domain of σ' and the functions agree on their common domain. This is an inductive ordering of $\mathcal{P}$, as we now explain. Assume that $(\sigma^\mu)_{\mu<\nu}$ is an increasing sequence of sequences in $\mathcal{P}$. Define Σ' to be the union of the functions σ^μ. Then Σ' is a function from the ordinal $\rho := \bigcup_{\mu<\nu} \mathrm{dom}(\sigma^\mu)$ into $\mathbf{Con}(\mathbf{A})$ that preserves all existing joins. Σ' may not be in $\mathcal{P}$ because elements of $\mathcal{P}$ are functions from successor ordinals into $\mathbf{Con}(\mathbf{A})$ and ρ may be a limit ordinal. But if it is, then there is a unique meet continuous extension of Σ' to $\rho + 1$: it is the sequence $\Sigma = (\Sigma_\lambda)_{\lambda<\rho+1}$ where $(\Sigma_\lambda)_{\lambda<\rho} = \Sigma'$ and Σ_ρ is defined to be $\bigcup_{\lambda<\rho} \Sigma_\lambda$. The facts that Σ is a meet continuous homomorphism from a successor ordinal into $\mathbf{Con}(\mathbf{A})$, Σ extends each σ^μ, and $\Sigma_\lambda = \sigma^\mu_\lambda \lhd \sigma^\mu_{\lambda+1} = \Sigma_{\lambda+1}$ for some μ for each $\lambda < \rho$ imply that Σ is an upper bound in $\mathcal{P}$ for all the sequences σ^μ.

Since $\mathcal{P}$ is nonempty and inductively ordered, Zorn's lemma applies to show that $\mathcal{P}$ has a maximal element $(\psi_\lambda)_{\lambda<\kappa+1}$. The final congruence in this sequence,

ψ_κ, is ∞-solvable over Δ and is a maximal congruence in the interval $I[\Delta, \Theta]$ among those that are ∞-solvable over Δ. For each $i \in I$ we have $\delta_i \lll \theta_i$ and $\delta_i \leq \Delta \leq \psi_\kappa$, so by Lemma 6.10,

$$\psi_\kappa = (\psi_\kappa \vee \delta_i) \lll (\psi_\kappa \vee \theta_i) \leq \Theta\,.$$

The maximality condition on ψ_κ implies that $\psi_\kappa = \psi_\kappa \vee \theta_i$, or $\psi_\kappa \geq \theta_i$. Since this is true for all $i \in I$, we get that $\psi_\kappa \geq \bigvee_{i\in I} \theta_i = \Theta$. This proves that $\Delta \lll \psi_\kappa = \Theta$, as desired. □

DEFINITION 6.12. Assume that $\mathcal{V}$ is a variety with a weak difference term, and $\mathbf{A} \in \mathcal{V}$. If $\alpha \in \mathrm{Con}(\mathbf{A})$, then

$$\mathrm{rad}(\alpha) := \bigvee\{\beta \in \mathrm{Con}(\mathbf{A}) \mid \alpha \overset{s}{\sim} \beta\}$$

is called the **∞-solvable radical of** α. The **∞-solvable radical of A** is rad(0). A **radical congruence** is a congruence α of $\mathbf{A}$ such that $\mathrm{rad}(\alpha) = \alpha$.

THEOREM 6.13. *Let $\mathcal{V}$ be a variety with a weak difference term. Assume that $\mathbf{A} \in \mathcal{V}$ and that $\alpha, \beta \in \mathrm{Con}(\mathbf{A})$. Then*

(1) *$\mathrm{rad}(\alpha)$ is the largest congruence $\overset{s}{\sim}$-related to α.*
(2) $\mathrm{rad}\big(\mathrm{rad}(\alpha)\big) = \mathrm{rad}(\alpha)$.
(3) $\mathrm{rad}(\alpha \wedge \beta) = \mathrm{rad}(\alpha) \wedge \mathrm{rad}(\beta)$.
(4) *$\alpha \blacktriangleleft \beta$ if and only if $\alpha < \beta$ and $\alpha = \beta \wedge \mathrm{rad}(\alpha)$.*
(5) *If $\mathrm{Rad}(\mathbf{A})$ denotes the meet subsemilattice of $\mathbf{Con}(\mathbf{A})$ consisting of the radical congruences, then the restriction of the natural homomorphism*
$$\nu : \mathbf{Con}(\mathbf{A}) \to \ \mathbf{Con}(\mathbf{A})/\overset{s}{\sim} : \alpha \mapsto \alpha/\overset{s}{\sim}$$
to $\mathrm{Rad}(\mathbf{A})$ is an isomorphism onto $\mathbf{Con}(\mathbf{A})/\overset{s}{\sim}$.
(6) *The natural homomorphism ν from item (5) is meet continuous, and its image $\mathbf{Con}(\mathbf{A})/\overset{s}{\sim}$ is meet semidistributive.*

PROOF. For item (1), the fact that $\overset{s}{\sim}$ is compatible with the complete join operation implies that the join of all congruences $\overset{s}{\sim}$-related to α is $\overset{s}{\sim}$-related to α, and it is necessarily the largest congruence $\overset{s}{\sim}$-related to α.

For (2), $\alpha \overset{s}{\sim} \mathrm{rad}(\alpha)$ and $\mathrm{rad}(\alpha) \overset{s}{\sim} \mathrm{rad}\big(\mathrm{rad}(\alpha)\big)$ hold by (1), so $\alpha \overset{s}{\sim} \mathrm{rad}\big(\mathrm{rad}(\alpha)\big)$. Since $\mathrm{rad}(\alpha)$ is the largest congruence $\overset{s}{\sim}$-related to α, it follows that $\mathrm{rad}\big(\mathrm{rad}(\alpha)\big) \leq \mathrm{rad}(\alpha)$. The reverse inclusion follows from the facts that $\mathrm{rad}(\alpha) \overset{s}{\sim} \mathrm{rad}(\alpha)$ and $\mathrm{rad}\big(\mathrm{rad}(\alpha)\big)$ is the largest congruence $\overset{s}{\sim}$-related to $\mathrm{rad}(\alpha)$.

For (3), $\alpha \overset{s}{\sim} \mathrm{rad}(\alpha)$ and $\beta \overset{s}{\sim} \mathrm{rad}(\beta)$, so $\alpha \wedge \beta \overset{s}{\sim} \mathrm{rad}(\alpha) \wedge \mathrm{rad}(\beta)$ since $\overset{s}{\sim}$ is a congruence. Since $\mathrm{rad}(\alpha \wedge \beta)$ is the largest congruence $\overset{s}{\sim}$-related to $\alpha \wedge \beta$ we get that $\mathrm{rad}(\alpha) \wedge \mathrm{rad}(\beta) \leq \mathrm{rad}(\alpha \wedge \beta)$. On the other hand, $\alpha \wedge \beta \overset{s}{\sim} \mathrm{rad}(\alpha \wedge \beta)$, so

$$\alpha = \alpha \vee (\alpha \wedge \beta) \overset{s}{\sim} \alpha \vee \mathrm{rad}(\alpha \wedge \beta) \leq \mathrm{rad}(\alpha)$$

from the definition of $\mathrm{rad}(\alpha)$. Thus, $\mathrm{rad}(\alpha \wedge \beta) \leq \mathrm{rad}(\alpha)$, and similarly $\mathrm{rad}(\alpha \wedge \beta) \leq \mathrm{rad}(\beta)$, so $\mathrm{rad}(\alpha \wedge \beta) \leq \mathrm{rad}(\alpha) \wedge \mathrm{rad}(\beta)$. This completes the proof of (3).

For (4), assume that $\alpha \blacktriangleleft \beta$. Then $\alpha < \beta$ according to the definition of $\blacktriangleleft$. The congruence $\alpha' := \beta \wedge \mathrm{rad}(\alpha)$ satisfies $\alpha \leq \alpha' \leq \mathrm{rad}(\alpha)$ and is in $I[\alpha, \beta]$, so it is a congruence in $I[\alpha, \beta]$ that is ∞-solvable over α. If this fact is witnessed by a chain of congruences $(\theta_\lambda)_{\lambda<\kappa+1}$ starting at $\theta_0 = \alpha$ and ending at $\theta_\kappa = \alpha'$, and satisfying the conditions of Definition 6.5, then using the fact that $\alpha \blacktriangleleft \beta$ and induction it follows that $\theta_\lambda = \alpha$ for all λ, so $\alpha = \alpha' = \beta \wedge \mathrm{rad}(\alpha)$. Conversely, if $\alpha \blacktriangleleft \beta$ fails to hold and $\alpha < \beta$, then there is a congruence $\alpha' \in I[\alpha, \beta]$ such that $\alpha \lhd \alpha'$ and $\alpha < \alpha'$. Since $\alpha \overset{s}{\sim} \alpha'$ we get that $\alpha < \alpha' \leq \beta \wedge \mathrm{rad}(\alpha)$, so $\alpha = \beta \wedge \mathrm{rad}(\alpha)$ fails to hold.

For (5), it is clear that $\nu : \mathrm{Rad}(\mathbf{A}) \to \mathbf{Con}(\mathbf{A})/\overset{s}{\sim}$ is an order-preserving function between meet semilattices. To prove that it is an isomorphism it suffices to show that it is bijective, or equivalently that the elements of $\mathrm{Rad}(\mathbf{A})$ form a transversal for the congruence $\overset{s}{\sim}$. The fact that $\alpha \overset{s}{\sim} \mathrm{rad}(\alpha)$ (proved in (1)) together with the fact that $\mathrm{rad}(\alpha) \in \mathrm{Rad}(\mathbf{A})$ (an equivalent form of (2)) shows that every $\overset{s}{\sim}$-class contains a radical congruence. But we proved in (1) that every radical element is the largest element of its $\overset{s}{\sim}$-class, so each $\overset{s}{\sim}$-class has at most one radical congruence. This shows that the elements of $\mathrm{Rad}(\mathbf{A})$ form a transversal for $\overset{s}{\sim}$.

For (6), the fact that $\overset{s}{\sim}$ is compatible with the complete join operation implies that it is a meet continuous congruence, hence the quotient modulo this congruence is meet continuous and the natural homomorphism $\nu : \mathbf{Con}(\mathbf{A}) \to \mathbf{Con}(\mathbf{A})/\overset{s}{\sim}$ is a meet continuous homomorphism.

To prove that $\mathbf{Con}(\mathbf{A})/\overset{s}{\sim}$ is meet semidistributive, suppose that $(\alpha/\overset{s}{\sim}) \wedge (\beta/\overset{s}{\sim}) = (\alpha/\overset{s}{\sim}) \wedge (\gamma/\overset{s}{\sim}) = (\delta/\overset{s}{\sim})$. From item (5) we get that $\mathrm{rad}(\alpha) \wedge \mathrm{rad}(\beta) = \mathrm{rad}(\alpha) \wedge \mathrm{rad}(\gamma) = \mathrm{rad}(\delta)$ in $\mathrm{Rad}(\mathbf{A})$. Since $\mathrm{Rad}(\mathbf{A})$ is a meet subsemilattice of $\mathbf{Con}(\mathbf{A})$, these equalities hold in $\mathbf{Con}(\mathbf{A})$. Theorem 2.19 (8) implies that $\mathbf{C}\big(\mathrm{rad}(\beta), \mathrm{rad}(\alpha); \mathrm{rad}(\delta)\big)$ and $\mathbf{C}\big(\mathrm{rad}(\gamma), \mathrm{rad}(\alpha); \mathrm{rad}(\delta)\big)$ hold. Theorem 2.19 (5) now yields that $\mathbf{C}\big(\mathrm{rad}(\beta) \vee \mathrm{rad}(\gamma), \mathrm{rad}(\alpha); \mathrm{rad}(\delta)\big)$ holds. If

$$\theta := \mathrm{rad}(\alpha) \wedge \big(\mathrm{rad}(\beta) \vee \mathrm{rad}(\gamma)\big),$$

then $\theta \geq \mathrm{rad}(\delta)$ and $\mathbf{C}\big(\theta, \theta; \mathrm{rad}(\delta)\big)$ holds by Theorem 2.19 (1). This shows that $\mathrm{rad}(\delta) \lhd \theta$. But $\mathrm{rad}(\delta) \blacktriangleleft 1$ or else $\mathrm{rad}(\delta) = 1$, by items (2) and (4) of this theorem, and in either case we get

$$\mathrm{rad}(\delta) = \theta = \mathrm{rad}(\alpha) \wedge \big(\mathrm{rad}(\beta) \vee \mathrm{rad}(\gamma)\big).$$

Using the isomorphism proved in (5) once again, we conclude that $(\alpha/\overset{s}{\sim}) \wedge \big((\beta/\overset{s}{\sim}) \vee (\gamma/\overset{s}{\sim})\big) = (\delta/\overset{s}{\sim})$. This proves that $\mathbf{Con}(\mathbf{A})/\overset{s}{\sim}$ is meet semidistributive. □

THEOREM 6.14. *Let $\mathcal{V}$ be a variety with a weak difference term. If $\mathbf{A} \in \mathcal{V}$ has congruences $\delta \leq \theta$, then θ is ∞-solvable over δ if and only if the interval $I[\delta, \theta]$ contains no solvability obstruction.*

PROOF. Assume first that θ is ∞-solvable over δ, and that $I[\alpha, \beta]$ is an arbitrarily chosen but nontrivial subinterval of $I[\delta, \theta]$. Then $\delta = \delta \wedge \alpha \overset{s}{\sim} \theta \wedge \alpha = \alpha$, so $\mathrm{rad}(\alpha) = \mathrm{rad}(\delta) \geq \theta \geq \beta$. This shows that $\alpha < \beta = \beta \wedge \mathrm{rad}(\alpha)$. By item (4) of Theorem 6.13, $I[\alpha, \beta]$ is not a solvability obstruction.

Conversely, assume that $I[\delta, \theta]$ contains no solvability obstruction. Let $\alpha = \theta \wedge \mathrm{rad}(\delta)$. Then since $\delta \leq \alpha \leq \mathrm{rad}(\delta)$ we get that $\delta \overset{s}{\sim} \alpha$, and hence that $\mathrm{rad}(\alpha) = \mathrm{rad}(\delta)$. This means that $\alpha = \theta \wedge \mathrm{rad}(\alpha)$. According to item (4) of Theorem 6.13, either $\alpha \blacktriangleleft \theta$ or else $\alpha = \theta$. The first possibility is ruled out by the assumption that $I[\delta, \theta]$ contains no solvability obstruction, so it must be that $\alpha = \theta$. Since $\delta \overset{s}{\sim} \alpha = \theta$, the proof is complete. □

LEMMA 6.15. *Let $\mathcal{V}$ be a variety with a weak difference term. Assume that $\mathbf{A} \in \mathcal{V}$ has congruences γ, γ', δ and δ' such that $\gamma' \lhd \gamma$ and $\delta' \lhd \delta$. Then $\gamma \circ \delta \subseteq \gamma' \circ \delta \circ \gamma \circ \delta'$.*

PROOF. Choose any $(a, c) \in \gamma \circ \delta$. There is a $b \in A$ such that $a \equiv_\gamma b \equiv_\delta c$, so

$$a \equiv_{\gamma'} d(a, b, b) \equiv_\delta d(a, b, c) \equiv_\gamma d(b, b, c) \equiv_{\delta'} c$$

shows that $(a, c) \in \gamma' \circ \delta \circ \gamma \circ \delta'$. □

THEOREM 6.16. *Let $\mathcal{V}$ be a variety with a weak difference term. If $\mathbf{A} \in \mathcal{V}$, then $\stackrel{s}{\sim}$-classes of $\mathbf{Con}(\mathbf{A})$ consist of permuting congruences. In particular, these classes are modular sublattices.*

PROOF. We assume that the theorem is false and argue to a contradiction. If some $\stackrel{s}{\sim}$-class of $\mathbf{Con}(\mathbf{A})$ contains a pair of nonpermuting congruences, then there is a subinterval $I[\sigma, \tau]$ of that $\stackrel{s}{\sim}$-class that contains the two congruences. (For example, we could take σ to be the meet and τ to be the join of the two nonpermuting congruences.) Let A denote the ordered set of all congruences $\theta \in I[\sigma, \tau]$ such that every congruence in $I[\sigma, \theta]$ permutes with every congruence in $I[\sigma, \tau]$:

$$A = \{\theta \in I[\sigma, \tau] \mid \forall \theta' \in I[\sigma, \theta], \forall \psi \in I[\sigma, \tau]\ (\theta' \circ \psi = \psi \circ \theta')\}.$$

A is nonempty (since $\sigma \in A$) and it is closed under updirected unions, therefore A has maximal elements. Let α be a maximal element of A. It must be that $\alpha < \tau$ since otherwise all congruences in $I[\sigma, \tau]$ permute, contrary to assumption. Since $I[\sigma, \tau]$ is ∞-solvable and $\sigma \leq \alpha < \tau$, there is a congruence $\widehat{\alpha}$ such that $\alpha < \widehat{\alpha} \leq \tau$ and $\alpha \lhd \widehat{\alpha}$. The set

$$B = \{\theta \in I[\sigma, \tau] \mid \forall \theta' \in I[\sigma, \theta], \forall \psi \in I[\sigma, \widehat{\alpha}]\ (\theta' \circ \psi = \psi \circ \theta')\}$$

of all $\theta \in I[\sigma, \tau]$ such that every congruence in $I[\sigma, \theta]$ permutes with every congruence in $I[\sigma, \widehat{\alpha}]$ is also nonempty and closed under updirected unions, so B has a maximal element β. We cannot have $\beta = \tau$ since this would force $\widehat{\alpha} \in A$, contradicting the maximality of α. Since $\beta < \tau$, there is a congruence $\widehat{\beta}$ such that $\beta < \widehat{\beta} \leq \tau$ and $\beta \lhd \widehat{\beta}$.

By the maximality of β and the fact that $\beta < \widehat{\beta}$, there is some $\delta \leq \widehat{\beta}$ which fails to permute with some $\gamma \leq \widehat{\alpha}$. Every congruence below β permutes with every congruence below $\widehat{\alpha}$, so $\delta \not\leq \beta$. Every congruence below α permutes with every congruence in $I[\sigma, \tau]$, so $\gamma \not\leq \alpha$. Therefore $\gamma' = \alpha \wedge \gamma$ is strictly less than γ and $\delta' = \beta \wedge \delta$ is strictly less than δ. The four congruences γ, γ', δ and δ' have the following properties:

(i) $\gamma' < \gamma \leq \widehat{\alpha}$, $\gamma' \leq \alpha$;
(ii) $\delta' < \delta \leq \widehat{\beta}$, $\delta' \leq \beta$;
(iii) $\gamma' \lhd \gamma$, $\delta' \lhd \delta$; and
(iv) any two congruences in the four element set $\{\gamma, \gamma', \delta, \delta'\}$ permute, except $\gamma \circ \delta \not\subseteq \delta \circ \gamma$.

(The first, second and fourth items follow from the way we chose the congruences α, $\widehat{\alpha}$, β, $\widehat{\beta}$, γ, γ', δ, and δ'. The third item follows from Lemma 6.10 since, for example, $\alpha \lhd \widehat{\alpha}$ so $\gamma' = \alpha \wedge \gamma \lhd \widehat{\alpha} \wedge \gamma = \gamma$.) By item (iii) and Lemma 6.15 we have $\gamma \circ \delta \subseteq \gamma' \circ \delta \circ \gamma \circ \delta'$. But by items (i), (ii) and (iv) we have

$$\gamma' \circ \delta \circ \gamma \circ \delta' = \delta \circ \gamma' \circ \gamma \circ \delta' = \delta \circ \gamma \circ \delta' = \delta \circ \delta' \circ \gamma = \delta \circ \gamma.$$

Thus $\gamma \circ \delta \subseteq \delta \circ \gamma$, in contradiction with (iv). This is a contradiction to the assumption that $I[\sigma, \tau]$ contains a pair of nonpermuting congruences.

For the final statement of the theorem, recall that any lattice of permuting equivalence relations is modular. □

This result may be used to establish a new characterization of varieties with a weak difference term. In the following theorem, a lattice $\mathbf{L}$ has an $\mathrm{SD}_\wedge$/Modular **factorization** in a category $\mathcal{C}$ of lattices if there is a $\mathcal{C}$-morphism of $\mathbf{L}$ onto a meet semidistributive lattice whose kernel classes are modular.

THEOREM 6.17. *The following are equivalent for a variety $\mathcal{V}$.*

1. *$\mathcal{V}$ has a weak difference term.*
2. *For every $\mathbf{A} \in \mathcal{V}$, $\mathbf{Con}(\mathbf{A})$ has an* $\mathrm{SD}_\wedge$/Modular *factorization in the category of meet continuous lattices.*
3. *For every $\mathbf{A} \in \mathcal{V}$, $\mathbf{Con}(\mathbf{A})$ has an* $\mathrm{SD}_\wedge$/Modular *factorization in the category of ordinary lattices.*
4. *For every $\mathbf{A} \in \mathcal{V}$, each $\mathrm{SD}_\wedge$-failure in $\mathbf{Con}(\mathbf{A})$ is modular.*
5. *If $\mathbf{L}$ belongs to the meet continuous congruence variety of $\mathcal{V}$, then each $\mathrm{SD}_\wedge$-failure in $\mathbf{L}$ is modular.*

PROOF. If $\mathcal{V}$ has a weak difference term, then the natural homomorphism $\nu\colon \mathbf{Con}(\mathbf{A}) \to \mathbf{Con}(\mathbf{A})/\overset{s}{\sim}$ is a homomorphism of meet continuous lattices whose image is meet semidistributive according to Theorem 6.13 (6). The kernel classes are modular according to Theorem 6.16. This proves that (1) $\Longrightarrow$ (2).

The implications (2) $\Longrightarrow$ (3) $\Longrightarrow$ (4) are straightforward; the first since any factorization in the category of meet continuous lattices is a factorization in the category of lattices, and the second because all $\mathrm{SD}_\wedge$-failures must be collapsed by any homomorphism into a meet semidistributive lattice.

Assume that $\mathrm{SD}_\wedge$-failures of $\mathbf{Con}(\mathbf{A})$ are modular for any $\mathbf{A} \in \mathcal{V}$. As explained in the third paragraph of Remark 2.24, the class of meet continuous lattices whose $\mathrm{SD}_\wedge$-failures are modular is definable by a meet continuous identity. If (4) holds, then the congruence lattices of algebras in $\mathcal{V}$ satisfy this identity, so the meet continuous congruence variety of $\mathcal{V}$ also satisfies it, hence (5) holds.

By Theorem 4.7 the class of varieties $\mathcal{U}$ that satisfy the meet continuous congruence identity expressing the fact that $\mathrm{SD}_\wedge$-failures are modular is definable by idempotent Maltsev conditions. To show that (5) $\Longrightarrow$ (1) we must show that these Maltsev conditions are strong enough to force the existence of a weak difference term. By Theorem 6.2, it is enough to show that if $\mathcal{V}$ is a variety satisfying (5) then the abelian algebras in $\mathcal{V}$ are affine.

So let $\mathcal{V}$ be any variety satisfying condition (5), and let $\mathcal{A}$ be the subvariety of $\mathcal{V}$ that is generated by the abelian algebras of $\mathcal{V}$. Then $\mathcal{A}$ must also satisfy item (5), and in particular (since the variety of sets does not satisfy (5)) this implies that $\mathcal{A}$ has a Taylor term. Let $\mathbf{F} = \mathbf{F}_{\mathcal{A}}(4)$ be the 4-generated free algebra of this variety. The algebra $\mathbf{F}$ is abelian since $\mathcal{A}$ is generated by abelian algebras, the abelian property is preserved by subalgebras and products, and free algebras may be constructed as subalgebras of products of generating algebras. Since $\mathbf{F}$ is abelian, there is a congruence Δ on $\mathbf{F}^2$ that has the diagonal $D = \{(f, f) \mid f \in F\}$ as a class. It follows from Lemma 4.4 of [**53**] (in the case $\gamma = \delta = 1$ of that result) that since $\mathcal{A}$ has a Taylor term, the congruence Δ is a lattice complement of the coordinate projection kernels $\eta_1, \eta_2 \in \mathrm{Con}(\mathbf{F}^2)$. Since $\Delta \wedge \eta_1 = \Delta \wedge \eta_2 = 0$, the interval $I[0, \Delta \wedge (\eta_1 \vee \eta_2)] = I[0, \Delta]$ is an $\mathrm{SD}_\wedge$-failure. By item (5), this interval is modular. Since the diagonal D is a Δ-class, it follows from Lemma 2.4 of **[34]** that restriction is a lattice homomorphism from the interval $I[0, \Delta]$ to $\mathbf{Con}(\mathbf{F}^2|_D)$, where $\mathbf{F}^2|_D$ is the algebra induced on D by $\mathbf{F}^2$, so $\mathbf{Con}(\mathbf{F}^2|_D)$ is modular. The operations of $\mathbf{F}^2|_D$ are the polynomial operations of $\mathbf{F}^2$ under which D is closed, each of which agrees with some polynomial operation of $\mathbf{F}$ acting coordinatewise on D. Thus, $\mathbf{F}^2|_D$ is isomorphic as a nonindexed algebra to the algebra $\mathbf{F}|_F$ that $\mathbf{F}$ induces on itself. But the congruence lattices of $\mathbf{F}$ and $\mathbf{F}|_F$ are the same, so $\mathbf{F} = \mathbf{F}_\mathcal{A}(4)$ has a modular congruence lattice. By Day's Theorem, **[8]**, this implies that $\mathcal{A}$ is congruence modular. Since $\mathcal{A}$ is generated by abelian algebras, modular commutator theory implies that all algebras in $\mathcal{A}$ are affine. □

EXAMPLE 6.18. There exists a purely lattice-theoretic proof that (2) $\Longrightarrow$ (3) $\Longrightarrow$ (4) in Theorem 6.17, but not for the reverse implications. Those implications are false even for congruence lattices of individual algebras.

For example, if $\mathbf{L}$ is a lattice of height ≤ 3, then any $\mathrm{SD}_\wedge$-failure has height ≤ 2, hence is modular. Thus, any finite, nonmodular, simple lattice of height ≤ 3 has the properties that it is algebraic (hence isomorphic to the congruence lattice of some algebra), has modular $\mathrm{SD}_\wedge$-failures, but has no $\mathrm{SD}_\wedge$/Modular factorization. There are many graph lattices satisfying the required conditions (cf. Section 4.4).

There also exist algebraic lattices with $\mathrm{SD}_\wedge$/Modular factorizations in the category of ordinary lattices and no such factorization in the category of meet continuous lattices, such as the one in Figure 6.1.

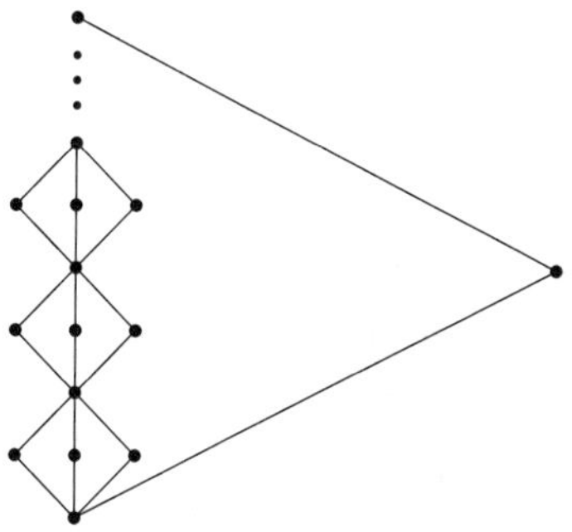

FIGURE 6.1. No $\mathrm{SD}_\wedge$/Modular factorization in $\mathcal{L}_{MC}$

We next show that the relation $\stackrel{s}{\sim}$ behaves well with respect to the class operators H, S and P_{fin}.

THEOREM 6.19. *Let $\mathcal{V}$ be a variety with a weak difference term.*

(1) *If $\mathbf{A} \in \mathcal{V}$ and $\alpha, \beta \geq \gamma$ are congruences on $\mathbf{A}$, then $\alpha \stackrel{s}{\sim} \beta$ in $\mathbf{Con}(\mathbf{A})$ if and only if $\alpha/\gamma \stackrel{s}{\sim} \beta/\gamma$ in $\mathbf{Con}(\mathbf{A}/\gamma)$.*
(2) *If $\mathbf{B}$ is a subalgebra of $\mathbf{A} \in \mathcal{V}$, then $\alpha \stackrel{s}{\sim} \beta$ in $\mathbf{Con}(\mathbf{A})$ implies that $\alpha|_B \stackrel{s}{\sim} \beta|_B$ in $\mathbf{Con}(\mathbf{B})$.*
(3) *If $\mathbf{A} \in \mathcal{V}$ has congruences α and β and $\mathbf{B} \in \mathcal{V}$ has congruences γ and δ, then $\alpha \times \gamma \stackrel{s}{\sim} \beta \times \delta$ in $\mathbf{Con}(\mathbf{A} \times \mathbf{B})$ if and only if $\alpha \stackrel{s}{\sim} \beta$ in $\mathbf{Con}(\mathbf{A})$ and $\gamma \stackrel{s}{\sim} \delta$ in $\mathbf{Con}(\mathbf{B})$.*

In particular, the class of ∞-solvable algebras in $\mathcal{V}$ is closed under the formation of homomorphic images, subalgebras and finite products.

PROOF. It follows from Theorem 2.19 (10) that if $\gamma \leq \delta \leq \theta$, then $\delta \lhd \theta$ if and only if $\delta/\gamma \lhd \theta/\gamma$. It is also true that a sequence $(\theta_\lambda)_{\lambda<\kappa}$ of congruences on $\mathbf{A}$, each containing γ, is continuous if and only if $(\theta_\lambda/\gamma)_{\lambda<\kappa}$ is. The property $\alpha \overset{s}{\sim} \beta$ in $\mathbf{Con}(\mathbf{A})$ is the property that there exists a continuous sequence of congruences witnessing ∞-solvability from $\alpha \wedge \beta$ to $\alpha \vee \beta$, which is therefore equivalent to the existence of a continuous sequence of congruences witnessing ∞-solvability from $(\alpha \wedge \beta)/\gamma = (\alpha/\gamma) \wedge (\beta/\gamma)$ to $(\alpha \vee \beta)/\gamma = (\alpha/\gamma) \vee (\beta/\gamma)$. Hence $\alpha \overset{s}{\sim} \beta$ is equivalent to $\alpha/\gamma \overset{s}{\sim} \beta/\gamma$ when $\gamma \leq \alpha \wedge \beta$. This proves (1).

It follows from Theorem 2.19 (9) that restriction to $\mathbf{B}$ preserves the relation $\lhd$ (i.e., $\delta \lhd \theta$ in $\mathbf{Con}(\mathbf{A})$ implies $\delta|_{\mathbf{B}} \lhd \theta|_{\mathbf{B}}$). Since restriction to $\mathbf{B}$ also preserves unions of chains, it must preserve $\lhd\!\!\lhd$. Hence ∞-solvable intervals in $\mathbf{Con}(\mathbf{A})$ restrict to ∞-solvable intervals in $\mathbf{Con}(\mathbf{B})$. Since $\alpha \overset{s}{\sim} \beta$ if and only if α and β lie in a single ∞-solvable interval, $\overset{s}{\sim}$ is also preserved under restriction to $\mathbf{B}$. This proves (2).

Applying item (1) to $\mathbf{A} \times \mathbf{B}$ (with the coordinate projection kernels η_1 and η_2 playing the role of γ in (1)) we obtain that $\alpha \overset{s}{\sim} \beta$ in $\mathbf{Con}(\mathbf{A})$ if and only if $\alpha_1 \overset{s}{\sim} \beta_1$ in $\mathbf{Con}(\mathbf{A}\times\mathbf{B})$ and similarly $\gamma \overset{s}{\sim} \delta$ in $\mathbf{Con}(\mathbf{B})$ if and only if $\gamma_2 \overset{s}{\sim} \delta_2$ in $\mathbf{Con}(\mathbf{A}\times\mathbf{B})$. Since ∞-solvability is a congruence, the conjunction of $\alpha \overset{s}{\sim} \beta$ and $\gamma \overset{s}{\sim} \delta$ implies that

$$\alpha \times \gamma := \alpha_1 \wedge \gamma_2 \overset{s}{\sim} \beta_1 \wedge \delta_2 =: \beta \times \delta\,.$$

Conversely, if $\alpha \times \gamma \overset{s}{\sim} \beta \times \delta$, then again because ∞-solvability is a congruence we get

$$\alpha_1 = \eta_1 \vee (\alpha \times \gamma) \overset{s}{\sim} \eta_1 \vee (\beta \times \delta) = \beta_1\,,$$

and similarly $\gamma_2 \overset{s}{\sim} \delta_2$. As proved above, these instances of $\overset{s}{\sim}$ hold if and only if $\alpha \overset{s}{\sim} \beta$ in $\mathbf{Con}(\mathbf{A})$ and $\gamma \overset{s}{\sim} \delta$ in $\mathbf{Con}(\mathbf{B})$. □

By a **partial Maltsev operation** on a set A we mean a function $m\colon D \to A$, for some domain $D \subseteq A^3$ containing all triples (x, y, y) and (y, y, x), $x, y \in A$, which satisfies $m(x, y, y) = x = m(y, y, x)$ for all $x, y \in A$.

THEOREM 6.20. *Let $\mathcal{V}$ be a variety with a weak difference term. Assume that $\mathbf{A} \in \mathcal{V}$, $\theta \in \mathrm{Con}(\mathbf{A})$ is ∞-solvable, and $S \subseteq A^n[\theta]$ is a subset. If S is compatible with the operations of $\mathbf{A}$, then S is also compatible with some partial Maltsev operation on A.*

PROOF. Recall from the opening paragraph of Chapter 2 that $A^n[\theta]$ is defined to be $\{\mathbf{x} \in A^n \mid x_i \equiv_\theta x_j \text{ for all } i, j\}$. Assume that $S \subseteq A^n[\theta]$ is compatible with the operations of $\mathbf{A}$. We will show that S is compatible with the partial Maltsev operation on A with smallest domain, which is

$$D = \{(a, b, b) \mid a, b \in A\} \cup \{(a, a, b) \mid a, b \in A\}\,.$$

We must show that if $\mathbf{a}, \mathbf{b}, \mathbf{c} \in S$ and the tuple

$$m(\mathbf{a}, \mathbf{b}, \mathbf{c}) = \big(m(a_1, b_1, c_1), m(a_2, b_2, c_2), \ldots, m(a_n, b_n, c_n)\big) =: \mathbf{d}$$

is defined in all coordinates, then $\mathbf{d} \in S$. To be defined in all coordinates, it must be the case that $a_i = b_i$ or $b_i = c_i$ (or both) for all i. Assume that this is so. Let $I \subseteq N := \{1, \ldots, n\}$ be the set of coordinates where $a_i = b_i$. If $I = \emptyset$, then $\mathbf{b} = \mathbf{c}$, so $\mathbf{d} = \mathbf{a} \in S$. If $I = N$, then $\mathbf{a} = \mathbf{b}$, so $\mathbf{d} = \mathbf{c} \in S$. Thus the only nontrivial cases are those where $\emptyset \neq I \neq N$. Restrict attention to those, and set $J = N - I$.

Let $\mathbf{S}$ be the subalgebra of $\mathbf{A}^n$ that is supported by $S = A^n[\theta]$. Elements $\mathbf{u}, \mathbf{v} \in S$ are related by θ_i $(= \pi_i^{-1}(\theta))$ if and only if $u_i \equiv_\theta v_i$. Since $S \subseteq A^n[\theta]$ this implies that $u_j \equiv_\theta u_i \equiv_\theta v_i \equiv_\theta v_j$ for any j, hence that $\theta_i \subseteq \theta_j$ for any i and j. Let $\widehat{\theta}$ denote the congruence $\theta_1 = \theta_2 = \cdots = \theta_n$ of $\mathbf{S}$. It follows from Theorem 6.19 (1) and (2) that $\theta_i \overset{s}{\sim} \eta_i$ for each i, so $\widehat{\theta} = \bigwedge \theta_i \overset{s}{\sim} \bigwedge \eta_i = 0$, proving that $\widehat{\theta}$ is ∞-solvable.

If $i \in I$, then $\eta_I \leq \eta_i \leq \theta_i = \widehat{\theta}$, and a similar argument proves that $\eta_J \leq \widehat{\theta}$. Since $\widehat{\theta}$ is ∞-solvable, $\eta_I \overset{s}{\sim} \widehat{\theta} \overset{s}{\sim} \eta_J$. This and Theorem 6.16 prove that η_I and η_J are permuting congruences of $\mathbf{S}$. According to the definition of I and J, we have $\mathbf{a}\ \eta_I\ \mathbf{b}\ \eta_J\ \mathbf{c}$, so there must exist a tuple $\mathbf{x} \in S$ such that $\mathbf{a}\ \eta_J\ \mathbf{x}\ \eta_I\ \mathbf{c}$. The tuple $\mathbf{x}$ satisfies the condition that $x_i = c_i$ if $i \in I$ and $x_j = a_j$ if $j \in N - I = J$. That is, $x_i = c_i$ if $m(a_i, b_i, c_i) = m(a_i, a_i, c_i) = c_i$, and $x_j = a_j$ if $m(a_j, b_j, c_j) = m(a_j, c_j, c_j) = a_j$. This proves that $\mathbf{x} = \mathbf{d}$, hence $\mathbf{d} = \mathbf{x} \in S$ as desired. □

It is a general fact that if $\mathbf{A}$ is an algebra, m is an operation on A, and for every cardinal κ any relation $S \subseteq A^\kappa$ compatible with the operations of $\mathbf{A}$ is also compatible with m, then m is a term operation of $\mathbf{A}$. If instead one only has that any *finitary* relation S on A that is compatible with the operations of $\mathbf{A}$ is also compatible with m, then m is only guaranteed to be a **local term operation** of $\mathbf{A}$. This means that for any finite subset $F \subseteq A$ there is a term operation t_F that agrees with m on F. Theorem 6.20 suggests the possibility that if $\mathcal{V}$ has a weak difference term and $\mathbf{A} \in \mathcal{V}$ is ∞-solvable, then $\mathbf{A}$ has a local Maltsev term operation. Admittedly Theorem 6.20 refers only to partial Maltsev operations, but that is exactly what one would expect of a result like Theorem 6.20 if the local Maltsev term operation were not unique. These observations lead to speculations recorded in the following problem.

PROBLEM 6.21. Assume that $\mathcal{V}$ has a weak difference term, and that $\mathbf{A} \in \mathcal{V}$ is ∞-solvable. Does $\mathbf{A}$ have a local Maltsev term operation? Does $\mathbf{A}$ have a Maltsev term operation? If $\mathcal{V}$ is generated by ∞-solvable algebras, then must $\mathcal{V}$ be congruence permutable?

Although affirmative answers to these questions seem unlikely, it is true that for each finite n there is a term that is Maltsev on classes of n-step solvable congruences of algebras in $\mathcal{V}$. It would be interesting if this property extended to congruences with higher degrees of solvability.

COROLLARY 6.22. *Let $\mathcal{V}$ be a variety with a weak difference term. If $\mathbf{A} \in \mathcal{V}$, $\rho \overset{s}{\sim} 0$ in $\mathbf{Con}(\mathbf{A})$ and $S \subseteq \rho$ $(= A^2[\rho])$ is a reflexive compatible binary relation on $\mathbf{A}$, then S is a congruence.*

PROOF. This is proved in the same way that one proves that a reflexive compatible binary relation on a Maltsev algebra is a congruence, since that argument works equally well with a partial Maltsev operation.

For the record, the details are as follows. If $S \subseteq \rho$ is reflexive and compatible, then S is compatible with a partial Maltsev operation according to Theorem 6.20. Thus, if $(a, b) \in S$, then since S is reflexive $(a, a), (b, b) \in S$ also, so $(b, a) = m\big((a, a), (a, b), (b, b)\big) \in S$. This proves that S is symmetric. Suppose that $(a, b), (b, c) \in S$, then $(a, c) = m\big((a, b), (b, b), (b, c)\big) \in S$. This proves that S is transitive, hence a congruence. □

THEOREM 6.23. *Let $\mathcal{V}$ be a variety with a weak difference term.*

(1) *If $\mathbf{A} \in \mathcal{V}$ has congruences $\alpha' \lll \alpha$ and $\beta' \lll \beta$, then*

$$\alpha \vee \beta = (\alpha' \vee \beta') \circ \alpha \circ \beta \circ (\alpha' \vee \beta').$$

(2) *If $\mathbf{A} \in \mathcal{V}$ has congruences α and β with $\beta \overset{s}{\sim} \alpha \wedge \beta$, then*

$$\alpha \vee \beta = \alpha \circ \beta \circ \alpha.$$

PROOF. For item (1), the relation $R = (\alpha' \vee \beta') \circ \alpha \circ \beta \circ (\alpha' \vee \beta')$ is an $(\alpha' \vee \beta')$-closed compatible reflexive relation on $\mathbf{A}$ whose transitive closure is $\alpha \vee \beta$. To prove this part of the theorem it suffices to show that R is a congruence.

Since $\alpha' \lll \alpha$ and $\beta' \lll \beta$, it follows from Lemma 6.10 and the transitivity of $\lll$ that $(\alpha' \vee \beta') \lll (\alpha \vee \beta)$. In the quotient algebra $\mathbf{A}/(\alpha' \vee \beta')$ we have that

$$\rho := (\alpha \vee \beta)/(\alpha' \vee \beta') \overset{s}{\sim} (\alpha' \vee \beta')/(\alpha' \vee \beta') = 0$$

by Theorem 6.19. The relation $S := R/(\alpha' \vee \beta')$ is a compatible reflexive relation of $\mathbf{A}/(\alpha' \vee \beta')$, and $S \subseteq \rho$, hence S is a congruence on $\mathbf{A}/(\alpha' \vee \beta')$ according to Corollary 6.22. But R is $(\alpha' \vee \beta')$-closed, so for the natural homomorphism $\nu : \mathbf{A} \to \mathbf{A}/(\alpha' \vee \beta') : a \mapsto a/(\alpha' \vee \beta')$ we have $R = \nu^{-1}(\nu(R))$. We know that $\nu(R) = R/(\alpha' \vee \beta') = S$ is a congruence, and the inverse image of a congruence is a congruence. Hence R is a congruence.

Item (2) is the special case of (1) where $\alpha = \alpha' \geq \beta' = \alpha \wedge \beta$. □

Theorem 6.16 follows from Theorem 6.23 (1) when $\alpha' = \beta' = \alpha \wedge \beta$.

THEOREM 6.24. *Let $\mathcal{V}$ be a variety with a weak difference term. If $\mathbf{A} \in \mathcal{V}$ has congruences α and β with $\beta \overset{s}{\sim} \alpha \wedge \beta$ and $\alpha \leq \gamma \leq \alpha \vee \beta$, then $\gamma = \alpha \vee (\beta \wedge \gamma)$. Hence the map $J : I[\alpha \wedge \beta, \beta] \to I[\alpha, \alpha \vee \beta] : x \mapsto x \vee \alpha$ is surjective.*

PROOF. The hypotheses of Theorem 6.23 (2) are satisfied, therefore $\alpha \vee \beta = \alpha \circ \beta \circ \alpha$. Arbitrarily choose $(a, b) \in \gamma$. Since $\gamma \leq \alpha \vee \beta = \alpha \circ \beta \circ \alpha$, there are elements $u, v \in A$ such that

$$a \equiv_\alpha u \equiv_\beta v \equiv_\alpha b. \tag{6.5}$$

This means that $u \equiv_\alpha a \equiv_\gamma b \equiv_\alpha v$, so $(u, v) \in \beta \wedge (\alpha \vee \gamma) = \beta \wedge \gamma$. This allows us to strengthen the β-relation in line (6.5) to a $(\beta \wedge \gamma)$-relation, and in the strengthened form that line indicates that $(a, b) \in \alpha \vee (\beta \wedge \gamma)$. Since (a, b) was chosen arbitrarily, $\gamma \subseteq \alpha \vee (\beta \wedge \gamma)$. The reverse inclusion follows from the fact that $\alpha \leq \gamma$.

For the final statement, note that if $\gamma \in I[\alpha, \alpha \vee \beta]$, then $\beta \wedge \gamma$ is in the interval $I[\alpha \wedge \beta, \beta]$ and we have proved above that $\gamma = J(\beta \wedge \gamma)$. □

THEOREM 6.25. *Let $\mathcal{V}$ be a variety with a weak difference term. Assume that $\mathbf{A} \in \mathcal{V}$.*

(1) *If $\mathbf{N}_5$, with the labeling indicated in Figure 6.2, appears as a sublattice of $\mathbf{Con}(\mathbf{A})$, then $\beta \overset{s}{\nsim} \beta \wedge \alpha$.*
(2) *If $\mathbf{D}_2$ appears as a sublattice of $\mathbf{Con}(\mathbf{A})$, then no two distinct congruences in the sublattice are ∞-solvably related.*

PROOF. In item (1), we have $\alpha \leq \gamma \leq \alpha \vee \beta$. Therefore, if we had $\beta \overset{s}{\sim} \beta \wedge \alpha$, then all of the hypotheses of Theorem 6.24 would be met. This would force $\gamma = \alpha \vee (\beta \wedge \gamma)$. But $\alpha \vee (\beta \wedge \gamma) = \alpha \neq \gamma$, so we cannot have $\beta \overset{s}{\sim} \beta \wedge \alpha$.

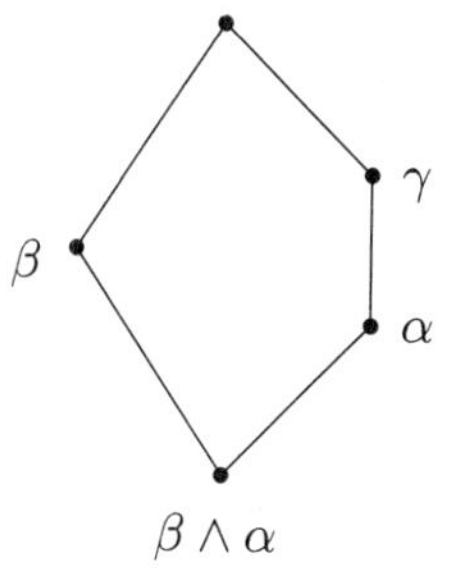

FIGURE 6.2. $\beta \overset{s}{\not\sim} \beta \wedge \alpha$

For item (2), the relation $\overset{s}{\sim}$ restricts to a congruence on $\mathbf{D}_2$, which is a subdirectly irreducible lattice. If the copy of $\mathbf{D}_2$ under consideration is labeled as in Figure 6.3, then the critical intervals are $I[\mu, \nu]$, $I[\alpha, \tau]$, and $I[\delta, \theta]$. Therefore, if any two distinct elements of this sublattice are $\overset{s}{\sim}$-related then $\mu \overset{s}{\sim} \nu$, $\alpha \overset{s}{\sim} \tau$, and $\delta \overset{s}{\sim} \theta$. Assume that this is so and choose α' so that $\alpha < \alpha' \leq \tau$ and $\alpha \lhd \alpha'$. By The-

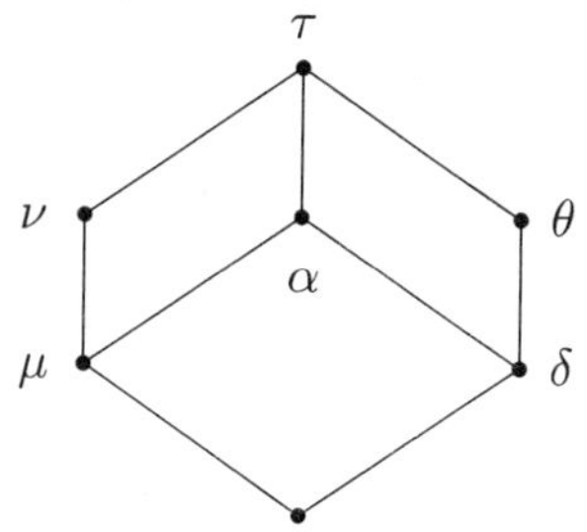

FIGURE 6.3

orem 6.24 the maps $I[\mu, \nu] \to I[\alpha, \tau] \colon x \mapsto x \vee \alpha$ and $I[\delta, \theta] \to I[\alpha, \tau] \colon x \mapsto x \vee \alpha$ are surjective, so there exist μ' and δ' such that $\mu < \mu' \leq \nu$, $\delta < \delta' \leq \theta$ and $\mu' \vee \alpha = \alpha' = \delta' \vee \alpha$. Since $I[\mu, \mu']$ and $I[\delta, \delta']$ are perspective with $I[\alpha, \alpha']$ we get that $\mu \lhd \mu'$ and $\delta \lhd \delta'$ from Theorem 3.27 (or Lemma 6.10).

CLAIM 6.26. *The congruences μ', α, and δ' and generate a sublattice isomorphic to $\mathbf{D}_2$.*

Our argument uses Lemma 5.27 with $x = \mu'$, $y = \alpha$ and $z = \delta'$, as well as the fact that ν, α and θ generate a copy of $\mathbf{D}_2$.

For item (i) of Lemma 5.27,

$$x \wedge z = \mu' \wedge \delta' \leq \nu \wedge \theta \leq \alpha = y \,.$$

For item (ii),

$$z \vee (y \wedge x) = \delta' \vee (\alpha \wedge \mu') = \delta' \vee \mu = \alpha' = \alpha \vee \mu' \geq \mu' = x \,.$$

Here we have used the facts that $I[\mu, \mu']$ and $I[\delta, \delta']$ are perspective up to $I[\alpha, \alpha']$, and that $\delta' \vee \mu = \delta' \vee (\delta \vee \mu) = \delta' \vee \alpha = \alpha'$.

Item (iii) is similar to (ii). For (iv),

$$(x \wedge y) \vee (y \wedge z) = (\mu' \wedge \alpha) \vee (\alpha \wedge \delta') = \mu \vee \delta = \alpha \,.$$

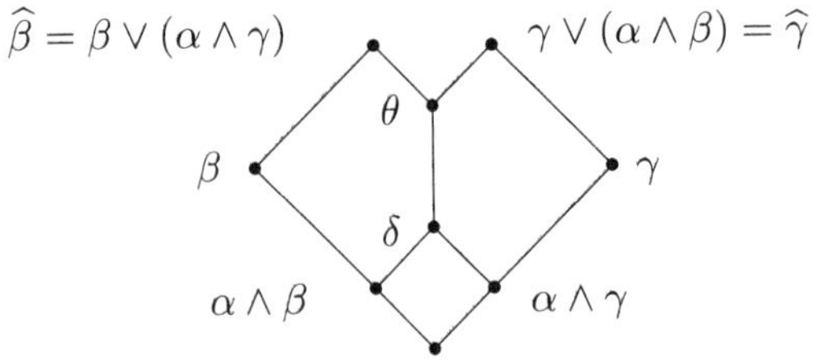

FIGURE 6.4

Since $x = \mu' \not\leq \alpha = y$, the sublattice generated by $\{\mu', \alpha, \delta'\}$ is isomorphic to $\mathbf{D}_2$. This proves the claim.

From the assumption that $\mathbf{Con(A)}$ has a sublattice isomorphic to $\mathbf{D}_2$ with ∞-solvable critical intervals we have shown that $\mathbf{Con(A)}$ has a sublattice isomorphic to $\mathbf{D}_2$ with abelian critical intervals. This contradicts Theorem 4.16 (2). □

The rest of the results in this section provide some information about failures of join semidistributivity in the congruence lattice of an algebra with a weak difference term.

LEMMA 6.27. *Assume that $\mathcal{V}$ has a weak difference term. If $\mathbf{A} \in \mathcal{V}$ has congruences α, β and γ satisfying*

(1) $\alpha \vee \beta \overset{s}{\sim} \alpha \overset{s}{\sim} \alpha \vee \gamma$, *and*
(2) $\beta \wedge \gamma \leq \alpha$,

then $\big(\beta \vee (\alpha \wedge \gamma)\big) \wedge \big(\gamma \vee (\alpha \wedge \beta)\big) = (\alpha \wedge \beta) \vee (\alpha \wedge \gamma)$.

PROOF. Since $\beta \wedge \gamma$ lies below all congruences of interest, we may factor by it and assume that $\beta \wedge \gamma = 0$.

In the next claim we label some of the congruences in the sublattice of $\mathbf{Con(A)}$ generated by $\{\alpha, \beta, \gamma\}$, and prove some of the more obvious relationships between them.

CLAIM 6.28. *Let* $\delta = (\alpha \wedge \beta) \vee (\alpha \wedge \gamma)$, $\widehat{\beta} = \beta \vee \delta$, $\widehat{\gamma} = \gamma \vee \delta$, $\theta = \widehat{\beta} \wedge \widehat{\gamma}$, $\beta' = \beta \wedge \theta$, *and* $\gamma' = \gamma \wedge \theta$. *Then*

(i) $\delta \leq \alpha$,
(ii) $\beta \wedge \delta = \beta \wedge \alpha$ *and* $\gamma \wedge \delta = \gamma \wedge \alpha$,
(iii) $\beta \overset{s}{\sim} \beta \wedge \delta$ *and* $\gamma \overset{s}{\sim} \gamma \wedge \delta$,
(iv) $\delta \leq \theta \leq \widehat{\beta}$ *and* $\delta \leq \theta \leq \widehat{\gamma}$,
(v) $\widehat{\beta} \overset{s}{\sim} \delta \overset{s}{\sim} \widehat{\gamma}$,
(vi) $\widehat{\beta} \vee \alpha = \beta \vee \alpha$ *and* $\widehat{\gamma} \vee \alpha = \gamma \vee \alpha$,
(vii) $\beta' \vee \delta = \theta = \gamma' \vee \delta$.

Some of these congruences and relationships are indicated in Figure 6.4.

We omit the proofs of the parts of the claim that concern γ if those parts are symmetric to the arguments we give for β.

For item (i) of Claim 6.28, note that $\delta = (\alpha \wedge \beta) \vee (\alpha \wedge \gamma)$ is a join of two elements below α, so $\delta \leq \alpha$.

For item (ii), since $\delta \leq \alpha$ we get that $\beta \wedge \delta \leq \beta \wedge \alpha$. On the other hand, both β and δ are above $\beta \wedge \alpha$, so $\beta \wedge \delta \geq \beta \wedge \alpha$.

For item (iii), start with $\alpha \overset{s}{\sim} \alpha \vee \beta$, which holds by assumption (1) of the lemma. Using the fact that $\overset{s}{\sim}$ is a congruence, meet both sides of $\alpha \overset{s}{\sim} \alpha \vee \beta$ with β to obtain $\beta \wedge \alpha \overset{s}{\sim} \beta$. Applying the result from (ii) of this claim yields $\beta \wedge \delta \overset{s}{\sim} \beta$.

For item (iv), the fact that $\theta = \widehat{\beta} \wedge \widehat{\gamma} = (\beta \vee \delta) \wedge (\gamma \vee \delta)$ is a meet of elements above δ is enough to prove that $\delta \leq \theta$, while $\theta \leq \widehat{\beta}$ is immediate from the definition $\theta = \widehat{\beta} \wedge \widehat{\gamma}$.

For item (v), join the relation $\beta \wedge \delta \overset{s}{\sim} \beta$ from (iii) with δ to obtain $\delta \overset{s}{\sim} \beta \vee \delta = \widehat{\beta}$.

For item (vi), $\widehat{\beta} \vee \alpha = (\beta \vee \delta) \vee \alpha = \beta \vee (\delta \vee \alpha) = \beta \vee \alpha$, where the last equality follows from item (i).

For item (vii), use the fact established in (iii) that $\beta \overset{s}{\sim} \beta \wedge \delta$, the fact established in (iv) that $\delta \leq \theta \leq \widehat{\beta} = \beta \vee \delta$, and Theorem 6.24 to deduce that $\theta = \delta \vee (\beta \wedge \theta)$. Since $\beta' = \beta \wedge \theta$, this yields $\theta = \beta' \vee \delta$. This completes the proof of Claim 6.28.

By examining the definitions of δ and θ, the reader will verify that the statement of the next claim is precisely what is asserted in the statement of this lemma.

CLAIM 6.29. $\delta = \theta$.

Assume instead that $\delta \neq \theta$. We show that the sublattice of $\mathbf{Con}(\mathbf{A})$ generated by $\{\beta', \gamma', \delta\}$ is isomorphic to $\mathbf{D}_2$. In order to show this, we verify that the congruences $x = \beta'$, $y = \delta$, and $z = \gamma'$ satisfy the relations from the presentation of $\mathbf{D}_2$ given in Lemma 5.27. Throughout the argument we will make free use of Claim 6.28.

To establish relation (i) of Lemma 5.27, we have

$$x \wedge z = \beta' \wedge \gamma' \leq \beta \wedge \gamma = 0 \leq \delta = y\,.$$

For relation (ii),

$$\begin{aligned} z \vee (y \wedge x) &= \gamma' \vee (\delta \wedge \beta') \\ &= (\gamma \wedge \theta) \vee \big(\delta \wedge (\beta \wedge \theta)\big) \\ &= \big(\gamma \wedge \theta\big) \vee \big(\beta \wedge \delta\big) \\ &= (\gamma \wedge \theta) \vee (\gamma \wedge \delta) \vee (\beta \wedge \delta) \\ &= \gamma' \vee \big((\gamma \wedge \alpha) \vee (\beta \wedge \alpha)\big) \\ &= \gamma' \vee \delta \\ &= \theta \geq \beta \wedge \theta = \beta' = x\,. \end{aligned}$$

The proof that $x \vee (y \wedge z) \geq z$ is similar to the one just given. For relation (iv),

$$\begin{aligned} (x \wedge y) \vee (y \wedge z) &= (\beta' \wedge \delta) \vee (\delta \wedge \gamma') \\ &= \big((\beta \wedge \theta) \wedge \delta\big) \vee \big(\delta \wedge (\gamma \wedge \theta)\big) \\ &= (\beta \wedge \delta) \vee (\gamma \wedge \delta) \\ &= (\beta \wedge \alpha) \vee (\gamma \wedge \alpha) \\ &= \delta = y\,. \end{aligned}$$

Finally, $x = \beta' \not\leq \delta = y$ since $\beta' \vee \delta = \theta$ by Claim 6.28 (vii), and we are assuming that $\delta \neq \theta$.

Thus, our assumption that $\delta \neq \theta$ and Lemma 5.27 together imply that the sublattice generated by $\{\beta', \gamma', \delta\}$ is isomorphic to $\mathbf{D}_2$. It follows from Claim 6.28 (iv) and (v) that $I[\delta, \theta]$ is an ∞-solvable interval in this sublattice. This is forbidden by Theorem 6.25, hence Claim 6.29 and the lemma are proved. □

THEOREM 6.30. *Assume that a variety $\mathcal{V}$ has a weak difference term. If $\mathbf{A} \in \mathcal{V}$ has congruences α, β and γ satisfying*

(1) $\alpha \vee \beta \overset{s}{\sim} \alpha \overset{s}{\sim} \alpha \vee \gamma$, *and*
(2) $\beta \wedge \gamma \leq \alpha$,

then $\mathbf{C}(\alpha \vee \beta, \alpha \vee \gamma; \alpha)$ *holds.*

PROOF. As in the proof of Lemma 6.27, we may factor by $\beta \wedge \gamma$, and thus assume that $\beta \wedge \gamma = 0$. As in that proof, let $\delta = (\alpha \wedge \beta) \vee (\alpha \wedge \gamma)$, $\widehat{\beta} = \beta \vee \delta$, and $\widehat{\gamma} = \gamma \vee \delta$.

CLAIM 6.31. $\alpha \wedge \widehat{\beta} = \alpha \wedge \widehat{\gamma} = \widehat{\beta} \wedge \widehat{\gamma}$.

We proved in Lemma 6.27 that $\delta = \widehat{\beta} \wedge \widehat{\gamma}$, so it suffices (by symmetry) to prove only that $\alpha \wedge \widehat{\beta} = \delta$. Both α and $\widehat{\beta}$ are above δ, according to the definitions of δ and $\widehat{\beta}$, so we certainly have $\alpha \wedge \widehat{\beta} \geq \delta$. This forces $\alpha \wedge \widehat{\beta} \in I[\delta, \widehat{\beta}] = I[\delta, \beta \vee \delta]$. The perspective interval $I[\beta \wedge \delta, \beta] = I[\beta \wedge \alpha, \beta]$ is ∞-solvable by assumption, so Theorem 6.24 guarantees that

$$\alpha \wedge \widehat{\beta} = \delta \vee \left(\beta \wedge (\alpha \wedge \widehat{\beta})\right).$$

But we can simplify the expression on the right as follows:

$$\delta \vee \left(\beta \wedge (\alpha \wedge \widehat{\beta})\right) = \delta \vee \left(\alpha \wedge (\beta \wedge \widehat{\beta})\right) = \delta \vee (\alpha \wedge \beta) = \delta\,.$$

This proves the claim.

CLAIM 6.32. $\mathbf{C}(\widehat{\beta}, \widehat{\gamma}; \alpha)$ *holds.*

It follows from Claim 6.31 that $\widehat{\gamma} \wedge \left(\widehat{\beta} \vee (\widehat{\gamma} \wedge \alpha)\right) \leq \alpha$, so Claim 6.32 is a direct application of Theorem 2.19 (8).

Now we complete the proof that $\mathbf{C}(\alpha \vee \beta, \alpha \vee \gamma; \alpha)$ holds. It follows from Theorem 2.19 (7) or (8) that $\mathbf{C}(\alpha, \widehat{\gamma}; \alpha)$ holds. From this, Claim 6.32, and Theorem 2.19 (5) we get that $\mathbf{C}(\alpha \vee \widehat{\beta}, \widehat{\gamma}; \alpha)$ holds. We established that $\alpha \vee \widehat{\beta} = \alpha \vee \beta$ in Claim 6.28 (vi) of Lemma 6.27, so $\mathbf{C}(\alpha \vee \beta, \widehat{\gamma}; \alpha)$ holds. Using Theorem 2.19 (3) we finally get that $\mathbf{C}(\alpha \vee \beta, \alpha \circ \widehat{\gamma} \circ \alpha; \alpha)$ holds. This claim, and this theorem, will be established if we show that $\alpha \circ \widehat{\gamma} \circ \alpha = \alpha \vee \widehat{\gamma}$ $(= \alpha \vee \gamma)$. For this, meet both sides of the relation $\alpha \vee \gamma \overset{s}{\sim} \alpha$, which is part of assumption (1) of the theorem, with the congruence $\widehat{\gamma}$. The result is that $\widehat{\gamma} \overset{s}{\sim} \alpha \wedge \widehat{\gamma}$. From Theorem 6.23 (2) we conclude that $\alpha \circ \widehat{\gamma} \circ \alpha = \alpha \vee \widehat{\gamma}$, as desired. □

There is a special case of the previous theorem worth recording.

COROLLARY 6.33. *Assume that $\mathcal{V}$ has a weak difference term, that $\mathbf{A} \in \mathcal{V}$, and that I is an* $\mathrm{SD}_\vee$*-failure in* $\mathbf{Con}(\mathbf{A})$. *If I is ∞-solvable, then I is abelian.*

PROOF. An $\mathrm{SD}_\vee$-failure is a nontrivial interval $I = I[\alpha, \alpha \vee \beta]$ determined by congruences α, β and γ satisfying $\alpha \vee \beta = \alpha \vee \gamma$ and $\beta \wedge \gamma \leq \alpha$. The assumption that I is ∞-solvable means that $\alpha \overset{s}{\sim} \alpha \vee \beta$ $(= \alpha \vee \gamma)$. This establishes hypothesis (1) of Theorem 6.30. Hypothesis (2) holds simply because I is an $\mathrm{SD}_\vee$-failure. Hence we have the conclusion of Theorem 6.30, that $\mathbf{C}(\alpha \vee \beta, \alpha \vee \gamma; \alpha)$ (equivalently $\mathbf{C}(\alpha \vee \beta, \alpha \vee \beta; \alpha)$) holds. This is precisely what it means for I to be abelian. □

6.3. An Alternative Development

In the preceding section we developed the theory of solvability in analogy with the theory of solvability for groups, using chains of congruences with abelian factors. In this short section we show that it is possible to develop the theory of solvability in analogy with the theory of solvability for rings. Namely, we show that the basic notions of the preceding section, rad, $\overset{s}{\sim}$ and $\lll$, can be defined in terms of prime congruences.

DEFINITION 6.34. Let $\mathbf{A}$ be an algebra. A congruence θ on $\mathbf{A}$ is **semiprime** if $\theta \blacktriangleleft 1$; it is **prime** if it is semiprime and meet irreducible in $\mathbf{Con}(\mathbf{A})$.

An m**-sequence** is a sequence $\sigma = (\theta_\lambda)_{\lambda<\omega}$ of principal congruences where $[\theta_\lambda, \theta_\lambda] \geq \theta_{\lambda+1}$ holds for all λ. The m-sequence σ **avoids** a congruence θ if $\theta_\lambda \not\leq \theta$ for all λ, otherwise it **intersects** θ.

Our definitions for "prime" and "semiprime" agree with those from ring theory. The concept of an m-sequence generalizes the concept of a multiplicatively closed subset of a commutative ring, or an m-system in a noncommutative ring. All of these notions are definable in terms of the centralizer relation.

Observe that if $(\theta_\lambda)_{\lambda<\omega}$ is an m-sequence, then it is descending, since $\theta_\lambda \geq [\theta_\lambda, \theta_\lambda] \geq \theta_{\lambda+1}$ for all λ.

THEOREM 6.35. *Let $\mathcal{V}$ be a variety with a weak difference term. Assume that $\mathbf{A} \in \mathcal{V}$, and $\theta \in \mathrm{Con}(\mathbf{A})$ is a proper congruence. The following conditions concerning θ are equivalent.*

(1) *θ is semiprime.*
(2) *θ is an intersection of prime congruences.*
(3) *For every principal congruence $\theta_0 \not\leq \theta$ there is an m-sequence $\sigma = (\theta_0, \theta_1, \ldots)$ starting at θ_0 that avoids θ.*
(4) *θ is a radical congruence.*

PROOF. It follows immediately from the definitions that (1) and (4) are equivalent, so we prove only that (1) $\Longrightarrow$ (3) $\Longrightarrow$ (2) $\Longrightarrow$ (1).

Assume that θ is semiprime and that $\theta_0 \not\leq \theta$. If $[\theta_0, \theta_0] \leq \theta$, then since $[\theta_0, \theta_0] \lhd \theta_0$ we get from Lemma 6.10 that

$$\theta = \theta \vee [\theta_0, \theta_0] \lhd \theta \vee \theta_0 .$$

But since $\theta < \theta \vee \theta_0$, this contradicts the assumption that $\theta \blacktriangleleft 1$ (i.e., that θ is semiprime). Thus it must be that $[\theta_0, \theta_0] \not\leq \theta$, and this means that there is a principal congruence $\theta_1 \leq [\theta_0, \theta_0]$ such that $\theta_1 \not\leq \theta$. The same argument applied to θ_1 produces a principal congruence θ_2 such that $\theta_2 \leq [\theta_1, \theta_1]$ and $\theta_2 \not\leq \theta$. Repeating the argument indefinitely produces an m-sequence $(\theta_0, \theta_1, \ldots)$ starting at θ_0 that avoids θ.

Now suppose that Condition (3) holds for θ. Let θ' be the (possibly empty) intersection of the prime congruences containing θ. If $\theta < \theta'$, then there is a principal congruence $\theta_0 \leq \theta'$ such that $\theta_0 \not\leq \theta$. By Condition (3), there is an m-sequence $\sigma = (\theta_0, \theta_1, \ldots)$ that avoids θ. Extend θ to a congruence π that is maximal for the property that σ avoids π. (The fact that such a congruence exists uses Zorn's Lemma, and relies on the fact that each θ_i in σ is compact.) We claim that π is a prime congruence. To see that it is meet irreducible, assume instead that $\pi = \alpha \wedge \beta$ where $\pi < \alpha, \beta$. Then by the maximality of π it must be that σ intersects both α

and β, so $\theta_i \leq \alpha$ and $\theta_j \leq \beta$ for some i and j. But then for $k = \max(i,j)$ we have $\theta_k \leq \alpha \wedge \beta = \pi$, since σ is a descending chain of congruences, and this contradicts the fact that σ avoids π. To show that π is semiprime, assume instead that there is a congruence δ such that $\pi \lhd \delta$ and $\pi < \delta$. Then the maximality condition on π forces σ to intersect δ, so for some i we have $\theta_i \leq \delta$. But now the monotonicity of the commutator operation implies that

$$\theta_{i+1} \leq [\theta_i, \theta_i] \leq [\delta, \delta] \leq \pi .$$

This contradicts the fact that σ avoids π. This concludes the proof that π is prime. Since π extends θ, and θ' is the intersection of prime congruences containing θ, it follows that $\theta' \leq \pi$. But now we have the contradiction that $\theta_0 \leq \theta' \leq \pi$ while $\sigma = (\theta_\lambda)_{\lambda<\omega}$ avoids π. This final contradiction proves that $\theta' = \theta$, and hence that Condition (2) holds.

Finally, assume that Condition (2) holds. Then $\theta = \bigcap_{i\in I} \pi_i$ where each π_i is a prime congruence. If $\theta \lhd \delta$, then (since $\theta \leq \pi_i$) we get $\pi_i \lhd \pi_i \vee \delta$ for each i. Since each π_i is semiprime we deduce that $\pi_i = \pi_i \vee \delta$, or that $\delta \leq \pi_i$ for all i. Hence $\delta \leq \bigcap_{i\in I} \pi_i = \theta$. This shows that $\theta \lhd \delta$ implies that $\theta = \delta$, so $\theta \blacktriangleleft 1$. Hence Condition (1) holds. □

The following corollary shows that the concepts introduced earlier in the chapter can be introduced via prime congruences.

Corollary 6.36. *Let $\mathcal{V}$ be a variety with a weak difference term. Assume that $\mathbf{A} \in \mathcal{V}$ has congruences α and β.*

(1) *$\mathrm{rad}(\alpha)$ equals the intersection of the prime congruences containing α.*
(2) *$\alpha \overset{s}{\sim} \beta$ if and only if α and β are contained in the same prime congruences.*
(3) *$\alpha \lll \beta$ if and only if $\alpha \leq \beta$ and every m-sequence that intersects β also intersects α.*

Proof. Item (1) follows from the fact, proved in Theorem 6.35, that the radical congruences are exactly the intersections of prime congruences. Item (2) follows from (1) and the fact that $\alpha \overset{s}{\sim} \beta$ if and only if $\mathrm{rad}(\alpha) = \mathrm{rad}(\beta)$.

For item (3), assume that $\alpha \lll \beta$ and that the m-sequence σ intersects β. If σ avoids α, then α can be extended to a prime π that σ avoids, as in the proof of Theorem 6.35 (3) $\Longrightarrow$ (2). The prime π cannot contain β since σ avoids π but intersects β. Therefore π contains α and not β. Since $\alpha \lll \beta$, this contradicts item (2) of this corollary. For the other direction, assume that $\alpha \not\lll \beta$. Then, since $\alpha \leq \beta$, item (2) guarantees that there is a prime π containing α and not containing β. If θ_0 is a principal congruence satisfying $\theta_0 \leq \beta$ and $\theta_0 \not\leq \pi$, then Theorem 6.35 guarantees the existence of an m-sequence starting at θ_0 and avoiding π since the prime π is semiprime. This m-sequence also avoids α since $\alpha \leq \pi$, but it does not avoid β since $\theta_0 \leq \beta$. Thus (3) holds. □

CHAPTER 7

Ordinary Congruence Identities

In this chapter we will prove that a variety satisfies a nontrivial lattice identity as a congruence identity if and only if it satisfies an idempotent Maltsev condition that fails in the variety of semilattices. In view of the results of Chapter 5, this shows that the class of varieties that satisfy a congruence identity is definable by a Maltsev condition (for example, the Maltsev condition of Theorem 5.28).

7.1. A Rank for Solvability Obstructions

The congruence identity that we describe depends on the number of variables in a Hobby–McKenzie term for the variety, so let us now fix that arity. We assume in this chapter that $\mathcal{V}$ is a variety with a fixed Hobby–McKenzie term $F(x_1, \ldots, x_n)$, and we let $N = \{1, 2, \ldots, n\}$. By Corollary 6.3, $\mathcal{V}$ also has a weak difference term, so we are free to use the results of Chapter 6 in this chapter.

DEFINITION 7.1. If $I[\delta, \theta]$ is a solvability obstruction and T is a tolerance, then T **separates** $I := I[\delta, \theta]$ if $T \subseteq \theta$ and $T \not\subseteq \delta$. We write $I[\delta, \theta]_T$ or I_T to refer to the obstruction I in a manner that indicates that T separates I.

If f and g are m-ary terms, then I_T **supports the local equation** $f(x_1, \ldots, x_m) \approx_{I_T} g(x_1, \ldots, x_m)$ if $f(x_1, \ldots, x_m) \equiv_\delta g(x_1, \ldots, x_m)$ holds whenever all $x_i \in \{a, b\}$ for each $(a, b) \in T$.

DEFINITION 7.2. Assume that $\mathbf{A} \in \mathcal{V}$, that $I = I[\delta, \theta]$ is a solvability obstruction, and that T is a tolerance that separates I. The **rank of** I_T is the set

$$\mathrm{Rank}(I_T) := \{U \subseteq N \mid F_U(x, y) \approx_{I_T} x\}\,.$$

LEMMA 7.3. *Assume that* $\mathbf{A} \in \mathcal{V}$ *and that* $I_T := I[\delta, \theta]_T$ *is a separated solvability obstruction in* $\mathbf{Con}(\mathbf{A})$. *Then*

(1) $N \in \mathrm{Rank}(I_T)$.
(2) $\emptyset \notin \mathrm{Rank}(I_T)$.
(3) $\mathrm{Rank}(I_T)$ *is a closed*[1] *subset of* $\mathcal{B}(F)$ *that is not a lattice filter.*

PROOF. For (1), $N \in \mathrm{Rank}(I_T)$ if and only if

$$F(x, x, \ldots, x) = F_N(x, y) \approx_{I_T} x\,.$$

That this is true follows from the fact that F is idempotent.

For (2), if $\emptyset \in \mathrm{Rank}(I_T)$, then

$$y = F(y, y, \ldots, y) = F_\emptyset(x, y) \approx_{I_T} x\,.$$

This statement means that $y \equiv_\delta x$ for all $(x, y) \in T$, contradicting the fact that T separates $I[\delta, \theta]$. Thus $\emptyset \notin \mathrm{Rank}(I_T)$.

[1] "Closed" was defined after Example 2.13.

To prove item (3), notice that if $\mathcal{V} \models F_U(x,y) \approx F_V(x,y)$, then $U \in \mathrm{Rank}(I_T)$ if and only if $V \in \mathrm{Rank}(I_T)$. Therefore $\mathrm{Rank}(I_T)$ is a closed subset of $\mathcal{B}(F)$. It is not a lattice filter since, by parts (2) and (3), it is proper and nonempty, and $\mathcal{B}(F)$ has no closed, proper, nonempty lattice filter. □

A certain configuration involving a pair of separated solvability obstructions forces one obstruction to have rank that properly contains the rank of the other obstruction. We identify this configuration now, and introduce a shorthand notation for it.

DEFINITION 7.4. $I[\mu,\nu]_S \rightsquigarrow I[\delta,\theta]_T$ means that

(a) $S \cap (\delta : T) \subseteq \mu$,
(b) $T \cap \big(S \circ (T \cap \delta) \circ (S \cap \mu)\big) \subseteq \delta$, and
(c) $S \cap \big((T \cap \delta) \circ (S \cap \mu) \circ (T \cap \delta)\big) \subseteq \mu$.

The three conditions of this definition describe a property of the tolerance

$$\overline{T} := T \cap \big(S \circ (T \cap \delta) \circ S\big).$$

This tolerance consists of all pairs (c,d) for which there exist elements a and b related as in the quadrangle in Figure 7.1. Condition (a) implies that $\mathbf{C}(S,T;\delta)$

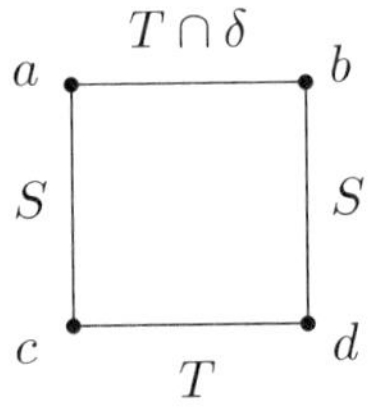

FIGURE 7.1

fails (since $S \not\subseteq \mu$). An S,T-matrix $\begin{bmatrix} a & b \\ c & d \end{bmatrix}$ witnessing this failure produces a pair $(c,d) \in \overline{T}$ with $(c,d) \notin \delta$. Thus, if Condition (a) holds, the tolerance $\overline{T} \subseteq T$ also separates $I[\delta,\theta]$. Condition (b) implies that if a, b, c and d are any four elements related as in the above quadrangle, and $(b,d) \in \mu$, then $(c,d) \in \delta$. If this happens, and Condition (c) holds, then $(a,c) \in \mu$. In particular, if $I[\mu,\nu]_S \rightsquigarrow I[\delta,\theta]_T$, then there will exist quadrangles related as above with $(c,d) \notin \delta$, but any such quadrangle must have $(a,c),(b,d) \notin \mu$.

LEMMA 7.5. *Assume that $I_S = I[\mu,\nu]_S$ and $J_T = I[\delta,\theta]_T$ are separated solvability obstructions in* $\mathbf{Con}(\mathbf{A})$, $\mathbf{A} \in \mathcal{V}$, *and that $I_S \rightsquigarrow J_T$. If $\overline{S} \subseteq S$ also separates $I[\mu,\nu]$ and $\overline{T} = T \cap \big(\overline{S} \circ (T \cap \delta) \circ \overline{S}\big)$, then*

(1) $I_{\overline{S}} \rightsquigarrow J_T$, *and*
(2) $\overline{T}$ *separates* $I[\delta,\theta]$.

PROOF. For item (1), suppose that $\overline{S} \subseteq S$ and that $\overline{S}$ separates I. Then the three conditions of the definition of $\rightsquigarrow$ hold with $\overline{S}$ in place of S, because the symbol S appears only on the left side of each inclusion and the left side of each inclusion is an expression that is monotone in S with respect to inclusion.

Item (2) follows from $I_{\overline{S}} \rightsquigarrow J_T$ as we explained in the remarks preceding the statement of this lemma (using S instead of $\overline{S}$). □

LEMMA 7.6. *Assume that $I_S = I[\mu,\nu]_S$ and $J_T = I[\delta,\theta]_T$ are separated solvability obstructions in* $\mathbf{Con}(\mathbf{A})$, $\mathbf{A} \in \mathcal{V}$, *and that $I_S \rightsquigarrow J_T$. Let $\overline{T} = T \cap \big(S \circ (T \cap \delta) \circ S\big)$. If $f(x,y,z)$ is an idempotent term of* $\mathbf{A}$, *and I_S supports the local equation $f(x,y,y) \approx_{I_S} x$, then both of the local equations*

$$f(x,x,y) \approx_{J_{\overline{T}}} x \quad \text{and} \quad f(x,y,x) \approx_{J_{\overline{T}}} x$$

are supported by $J_{\overline{T}}$.

PROOF. The fact that the tolerance $\overline{T} := T \cap \big(S \circ (T \cap \delta) \circ S\big)$ separates J follows from Condition (a) defining the relation $I_S \rightsquigarrow J_T$, as we remarked prior to the statement of Lemma 7.5 (or, it follows from Lemma 7.5 (2)). Now, to prove this lemma, it suffices to show that if I_S supports the local equation $f(x,y,y) \approx_{I_S} x$, then $J_{\overline{T}}$ supports the local equation $f(x,x,y) \approx_{J_{\overline{T}}} x$. The same argument applied to $f'(x,y,z) = f(x,z,y)$ shows that if the local equation $f(x,y,y) \approx_{I_S} x$ is supported by I_S, then $f(x,y,x) \approx_{J_{\overline{T}}} x$ is supported by $J_{\overline{T}}$.

To show that $J_{\overline{T}}$ supports the local equation $f(x,x,y) \approx_{J_{\overline{T}}} x$, choose a pair $(0,1) \in \overline{T} = T \cap \big(S \circ (T \cap \delta) \circ S\big)$. There exist elements u and v such that $0 \equiv_S u \equiv_{T\cap\delta} v \equiv_S 1$. The relationships between 0, 1, u and v are described in the left hand square in Figure 7.2. The right hand square of this figure describes the relationships between four other elements.

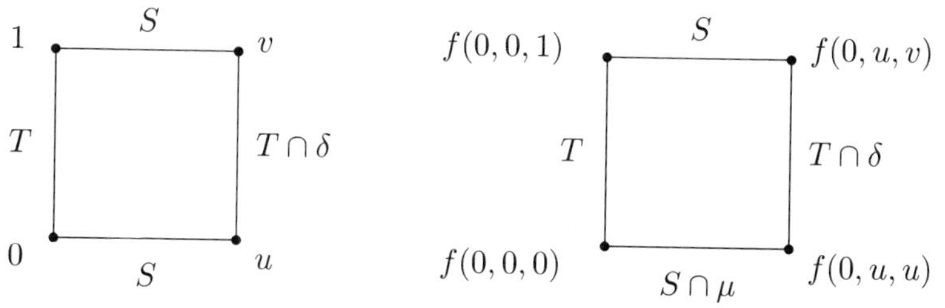

FIGURE 7.2

Let us convince ourselves that the information indicated in the right hand square of Figure 7.2 is correct. First,

$$f(0,0,0) \equiv_T f(0,0,1)$$

follows from the fact that $(0,0),(0,1) \in T$. Second,

$$f(0,u,u) \equiv_{T\cap\delta} f(0,u,v)$$

follows from the fact that $(0,0),(u,u),(u,v) \in T \cap \delta$. Third,

$$f(0,0,1) \equiv_S f(0,u,v) \quad \text{and} \quad f(0,0,0) \equiv_S f(0,u,u)$$

follow from $(0,0),(0,u),(1,v) \in S$. Finally,

$$f(0,0,0) \equiv_\mu f(0,u,u)$$

since both sides are μ-related to 0: the left hand side equals 0 since f is idempotent and the right hand side is μ-related to 0 since $(0,u) \in S$ and $I_S = I[\mu,\nu]_S$ supports $f(x,y,y) \approx_{I_S} x$.

The relations that we have just verified imply that

$$\big(f(0,0,1), f(0,0,0)\big) \in T \cap \big(S \circ (T \cap \delta) \circ (S \cap \mu)\big).$$

By Condition (b) of the definition of $I_S \rightsquigarrow J_T$ (Definition 7.4) this implies that

$$f(0,0,1) \equiv_\delta f(0,0,0) \ \ (=0)$$

holds for any pair $(0,1) \in \overline{T}$. Hence $J_{\overline{T}}$ supports the local equation $f(x,x,y) \approx_{J_{\overline{T}}} x$, and the proof is complete. □

LEMMA 7.7. *Assume that $I_S = I[\mu,\nu]_S$ and $J_T = I[\delta,\theta]_T$ are separated solvability obstructions in $\mathbf{Con}(\mathbf{A})$, $\mathbf{A} \in \mathcal{V}$, and that $I_S \rightsquigarrow J_T$. Let $\overline{T} = T \cap \big(S \circ (T\cap\delta) \circ S\big)$. If $f(x,y,z)$ is an idempotent term of $\mathbf{A}$, and I_S supports the local equations*

$$f(x,x,y) \approx_{I_S} x \quad \text{and} \quad f(x,y,x) \approx_{I_S} x\,,$$

then $f(x,y,y) \approx_{J_{\overline{T}}} x$ is supported by $J_{\overline{T}}$.

PROOF. As in the last proof, choose $(0,1) \in \overline{T} = T\cap\big(S\circ(T\cap\delta)\circ S\big)$, and let u and v denote elements of A for which $0 \equiv_S u \equiv_{T\cap\delta} v \equiv_S 1$. As before, relationships between 0, 1, u and v are expressed in Figure 7.3.

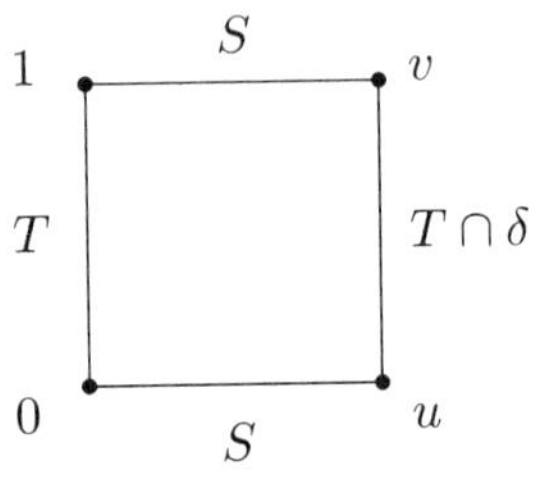

FIGURE 7.3

In this proof we will be interested in the relationships between some of the elements depicted in Figure 7.4. In this figure each element has the form $f(0,a,b)$.

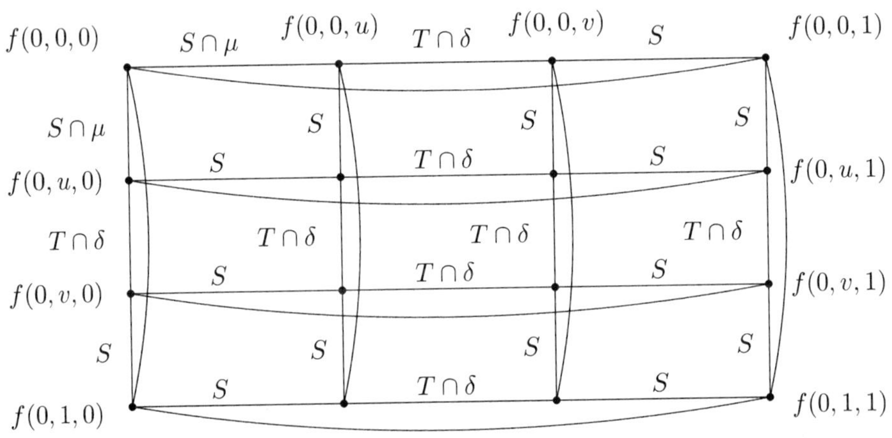

FIGURE 7.4

As one moves row-by-row from top to bottom the value of a changes from 0 to u to v to 1. As one moves column-by-column from left to right the value of b changes from 0 to u to v to 1. All but two of the indicated relations follow from $(0, u), (1, v) \in S$, $(u, v) \in T \cap \delta$, and that S and $T \cap \delta$ are compatible. The exceptions are the two instances of $S \cap \mu$ in the upper left hand corner of Figure 7.4. The relevant pairs are clearly S-related, since $(0, u) \in S$ and S is compatible, so we must justify the claim that

$$f(0, u, 0) \equiv_\mu f(0, 0, 0) \equiv_\mu f(0, 0, u) .$$

For this one can use the facts that $I_S = I[\mu, \nu]_S$ supports

$$f(x, y, x) \approx_{I_S} x = f(x, x, x) \approx_{I_S} f(x, x, y) ,$$

that $(0, u) \in S$, and that S is a tolerance. The curved lines in the figure represent T-relations, and all follow from $(0, 1) \in T$ and the fact that T is a tolerance.

We must prove that $f(0, 1, 1) \equiv_\delta 0$. We will accomplish this by showing that the curved lines of Figure 7.4 that appear across the top and along the right hand side represent $T \cap \delta$-relations, and not merely T-relations.

We begin our work by examining the top row of Figure 7.4. Following the straight line from right to left we see that

$$\big(f(0, 0, 1), f(0, 0, 0)\big) \in S \circ (T \cap \delta) \circ (S \cap \mu) ,$$

while following the curved path we see that these elements are related by T. By Condition (b) of the definition of $I_S \rightsquigarrow J_T$ we get that $\big(f(0, 0, 1), f(0, 0, 0)\big) \in \delta$. We can apply this information to the rectangle formed from the leftmost and rightmost elements of the first two rows, as shown in Figure 7.5. Condition (b) of the definition

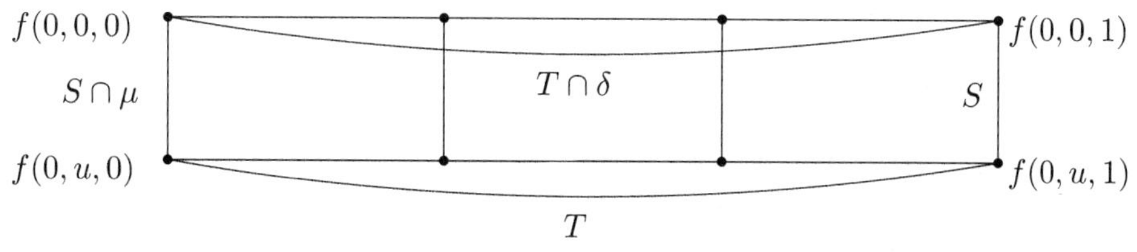

FIGURE 7.5

of $I_S \rightsquigarrow J_T$ can be applied again to show that

$$\big(f(0, u, 1), f(0, u, 0)\big) \in T \cap \big(S \circ (T \cap \delta) \circ (S \cap \mu)\big) \subseteq \delta ,$$

so the bottom curved line in this figure represents a δ-relation. Now we see that the two elements on the right hand side of this figure are related by $S \cap \big((T \cap \delta) \circ (S \cap \mu) \circ (T \cap \delta)\big)$. Condition (c) of the definition of $I_S \rightsquigarrow J_T$ asserts that $S \cap \big((T \cap \delta) \circ (S \cap \mu) \circ (T \cap \delta)\big) \subseteq \mu$, therefore $\big((f(0, 0, 1), f(0, u, 1)\big) \in \mu$.

Returning to Figure 7.4, we examine the four elements of the right hand side:

$$f(0, 0, 1) \equiv_{S \cap \mu} f(0, u, 1) \equiv_{T \cap \delta} f(0, v, 1) \equiv_S f(0, 1, 1) ,$$

and $f(0, 0, 1) \equiv_T f(0, 1, 1)$. Hence

$$\big(f(0, 1, 1), f(0, 0, 1)\big) \in T \cap \big(S \circ (T \cap \delta) \circ (S \cap \mu)\big) \subseteq \delta .$$

We have shown that the curved lines across the top and down the right side are δ-relations, so we have

$$f(0,0,0) \equiv_\delta f(0,0,1) \equiv_\delta f(0,1,1)\,.$$

Since f is idempotent, this proves that $f(0,1,1) \equiv_\delta f(0,0,0) = 0$, as desired. □

Next is the promised result showing that instances of $\rightsquigarrow$ are related to proper increases in rank.

THEOREM 7.8. *Assume that* $I_S = I[\mu,\nu]_S$ *and* $J_T = I[\delta,\theta]_T$ *are separated solvability obstructions in* $\mathbf{Con}(\mathbf{A})$, $\mathbf{A} \in \mathcal{V}$, *and* $I_S \rightsquigarrow J_T$. *If* $\overline{T} = T \cap \big(S \circ (T\cap\delta)\circ S\big)$, *then* $\mathrm{Rank}(I_S) \subsetneq \mathrm{Rank}(J_{\overline{T}})$. *In fact,*

(1) *if* $U \in \mathrm{Rank}(I_S)$ *and* $U \subseteq V \subseteq N$, *then* $V \in \mathrm{Rank}(J_{\overline{T}})$, *and*
(2) *if* $U, V \in \mathrm{Rank}(I_S)$ *and* $U \cup V = N$, *then* $U \cap V \in \mathrm{Rank}(J_{\overline{T}})$.

PROOF. It follows from Lemma 7.5 that $\overline{T}$ separates J, so $\mathrm{Rank}(J_{\overline{T}})$ is defined. We will first prove that (1) and (2) hold, and then we will deduce that $\mathrm{Rank}(I_S) \subsetneq \mathrm{Rank}(J_{\overline{T}})$. The following notation will be helpful: given a partition of $N = \{1,\ldots,n\}$ into three sets P, Q and R, one or two of which may be empty, we write $F_{P/Q/R}(x,y,z)$ to denote the term that results from substituting x for x_i in the Hobby–McKenzie term $F(x_1,\ldots,x_n)$ when $i \in P$, y for x_i when $i \in Q$, and z for x_i when $i \in R$. Any $F_{P/Q/R}(x,y,z)$ is idempotent, since F is.

To prove (1) assume that $U \in \mathrm{Rank}(I_S)$ and $U \subseteq V \subseteq N$. Partition N as $P/Q/R$ with $P = U$, $Q = V - U$ and $R = N - V$. The fact that $U \in \mathrm{Rank}(I_S)$ means precisely that the local equation

$$F_{U/(V-U)/(N-V)}(x,y,y) = F_U(x,y) \approx_{I_S} x$$

is supported by I_S. From Lemma 7.6 we get that $J_{\overline{T}}$ supports

$$F_V(x,y) = F_{U/(V-U)/(N-V)}(x,x,y) \approx_{J_{\overline{T}}} x\,.$$

Hence $V \in \mathrm{Rank}(J_{\overline{T}})$.

Now we prove (2). Assume that $U, V \in \mathrm{Rank}(I_S)$ and $U \cup V = N$. Partition N as $P/Q/R$ with $P = U\cap V$, $Q = U - V$ and $R = V - U$. The fact that $U, V \in \mathrm{Rank}(I_S)$ translates into

$$F_{(U\cap V)/(U-V)/(V-U)}(x,x,y) = F_U(x,y) \approx_{I_S} x$$

and

$$F_{(U\cap V)/(U-V)/(V-U)}(x,y,x) = F_V(x,y) \approx_{I_S} x\,,$$

so Lemma 7.7 proves that

$$F_{U\cap V}(x,y) = F_{(U\cap V)/(U-V)/(V-U)}(x,y,y) \approx_{J_{\overline{T}}} x$$

is supported by $J_{\overline{T}}$. Hence $U \cap V \in \mathrm{Rank}(J_{\overline{T}})$.

Now we use (1) and (2) to prove that $\mathrm{Rank}(I_S) \subsetneq \mathrm{Rank}(J_{\overline{T}})$. It follows from (1) that $U \in \mathrm{Rank}(I_S)$ implies $U \in \mathrm{Rank}(J_{\overline{T}})$, so we certainly have $\mathrm{Rank}(I_S) \subseteq \mathrm{Rank}(J_{\overline{T}})$. Suppose that $\mathrm{Rank}(I_S) = \mathrm{Rank}(J_{\overline{T}})$. Then from (1) we get that $\mathrm{Rank}(I_S)$ is an order filter in $\mathcal{B}(F)$. If $U, V \in \mathrm{Rank}(I_S)$ and $U \not\subseteq V$, then since V is the intersection of the $(n-1)$-element sets that contain it there must be an $(n-1)$-element set $W \supseteq V$ such that $U \not\subseteq W$. For this set W we have $W \in \mathrm{Rank}(I_S)$ and $U \cup W = N$, so by (2) we get that $U \cap W \in \mathrm{Rank}(J_{\overline{T}}) = \mathrm{Rank}(I_S)$. Since $U \cap W$ is a proper subset of U, and is in $\mathrm{Rank}(I_S)$, this shows that if U is not the least element of $\mathrm{Rank}(I_S)$, then U is not minimal under inclusion in $\mathrm{Rank}(I_S)$. It follows that $\mathrm{Rank}(I_S)$ is a principal order filter, hence a lattice filter in $\mathcal{B}(F)$. But by Lemma 7.3 (3) the set $\mathrm{Rank}(I_S)$ is not a lattice filter, hence $\mathrm{Rank}(I_S) = \mathrm{Rank}(J_{\overline{T}})$ is impossible. This completes the proof that $\mathrm{Rank}(I_S) \subsetneqq \mathrm{Rank}(J_{\overline{T}})$. □

Since an instance of the $\rightsquigarrow$-relation leads to a proper increase in rank, and there are only finitely many ranks, one expects a bound on the length of $\rightsquigarrow$-sequences. That bound is provided by the next theorem.

THEOREM 7.9. *Assume that* $\mathbf{A} \in \mathcal{V}$. *Then* $\mathbf{Con}(\mathbf{A})$ *does not contain a* $\rightsquigarrow$-*sequence of separated solvability obstructions*

$$I^0_{S^0} \rightsquigarrow I^1_{S^1} \rightsquigarrow I^2_{S^2} \rightsquigarrow \cdots \rightsquigarrow I^k_{S^k}$$

of length $k \geq 2n - 4$.

PROOF. Suppose otherwise, and that $I^j = I[\mu^j, \nu^j]$. Define tolerances $\overline{S}^j \subseteq S^j$ according to the rules

(i) $\overline{S}^0 := S^0$, and

(ii) $\overline{S}^{j+1} := S^{j+1} \cap (\overline{S}^j \circ (S^{j+1} \cap \mu^{j+1}) \circ \overline{S}^j)$.

Now consider two sequences of claims:

$(a)^j$ $I^j_{\overline{S}^j} \rightsquigarrow I^{j+1}_{S^{j+1}}$.

$(b)^j$ $\overline{S}^j$ separates I^j.

Claims $(a)^0$ and $(b)^0$ hold since $I^0_{S^0} \rightsquigarrow I^1_{S^1}$ and $\overline{S}^0 = S^0$ separates I^0. Claim $(a)^j$ implies Claim $(b)^{j+1}$ by Lemma 7.5 (2). Claim $(b)^j$ implies Claim $(a)^j$ by Lemma 7.5 (1) and the facts that $\overline{S}^j \subseteq S^j$ and $I^j_{S^j} \rightsquigarrow I^{j+1}_{S^{j+1}}$. By induction, Claims $(a)^j$ and $(b)^j$ hold for all j. Applying Theorem 7.8 to Claim $(a)^j$ yields that $\mathrm{Rank}(I^j_{\overline{S}^j}) \subsetneqq \mathrm{Rank}(I^{j+1}_{\overline{S}^{j+1}})$ for each $j < k$. This shows that

$$I^0_{S^0} \rightsquigarrow I^1_{S^1} \rightsquigarrow I^2_{S^2} \rightsquigarrow \cdots \rightsquigarrow I^k_{S^k}$$

implies that

$$\mathrm{Rank}(I^0_{\overline{S}^0}) \subsetneqq \mathrm{Rank}(I^1_{\overline{S}^1}) \subsetneqq \mathrm{Rank}(I^2_{\overline{S}^2}) \subsetneqq \cdots \subsetneqq \mathrm{Rank}(I^k_{\overline{S}^k})$$

where the tolerance $\overline{S}^j$ is contained in S^j yet still separates I^j.

Let $\mathcal{U}^j := \mathrm{Rank}(I^j_{\overline{S}^j})$. In the last paragraph we showed that

$$\mathcal{U}^0 \subsetneqq \mathcal{U}^1 \subsetneqq \mathcal{U}^2 \subsetneqq \cdots \subsetneqq \mathcal{U}^k.$$

Other properties of the sets $\mathcal{U}^j$ include:

- Each $\mathcal{U}^j$ is a closed subset of $\mathcal{B}(F)$ that is not a lattice filter (by Lemma 7.3 (3)).
- $N \in \mathcal{U}^j$ and $\emptyset \notin \mathcal{U}^j$ for all j (by Lemma 7.3 (1) and (2)).
- $\mathcal{U}^{j+1}$ contains the order filter generated by $\mathcal{U}^j$ (according to Lemma 7.8 (1)).
- $\mathcal{U}^{j+1}$ contains the intersections of comaximal pairs from $\mathcal{U}^j$ (by Lemma 7.8 (2)).

We now argue that such a sequence of sets cannot exist if $k \geq 2n - 4$. First, $\mathcal{U}^0$ contains N, is closed, and is not a lattice filter, so either $\mathcal{U}^0$ contains a set of size $\leq n - 2$ or else at least 2 sets of size $n - 1$. In either case, since $\mathcal{U}^1$

contains the order filter generated by $\mathcal{U}^0$ and contains the intersections of comaximal pairs in $\mathcal{U}^0$, the set $\mathcal{U}^1$ properly contains an order filter generated by a set U of size $n-2$. Now suppose that for some $r \geq 0$ the set $\mathcal{U}^{2r+1}$ contains a principal order filter generated by a set U of size $\leq n-2-r$. If U is not empty, then there is some $W \in \mathcal{U}^{2r+1}$ such that $W \not\supseteq U$ (since $\mathcal{U}^{2r+1}$ is closed). The set $V := U \cup (N - W)$ is in $\mathcal{U}^{2r+1}$ since $V \supseteq U$, and $V \cup W = N$ because $V \supseteq N - W$. Thus $V \cap W = U \cap W \in \mathcal{U}^{2r+2} = \mathcal{U}^{2(r+1)}$. Since $W \not\supseteq U$, the set $U \cap W$ has size $\leq n-2-(r+1)$. The set $\mathcal{U}^{2(r+1)+1}$ contains the order filter generated by $U \cap W$. By induction we see that for all r the set $\mathcal{U}^{2r+1}$ contains a principal order filter generated by a set of size $\leq n-2-r$. For $r = n-3$ we get that $\mathcal{U}^{2n-5}$ contains a principal order filter generated by a set U' of size 1. The containment is proper, since $\mathcal{U}^{2n-5}$ is not a lattice filter. If $W' \in \mathcal{U}^{2n-5}$ does not contain U', then as above we get that $V' := U' \cup (N - W') \in \mathcal{U}^{2n-5}$ and $V' \cap W' = \emptyset \in \mathcal{U}^{2n-4} \subseteq \mathcal{U}^k$. This is a contradiction, since no $\mathcal{U}^j$ contains the empty set. This completes the proof that there is no sequence

$$I^0_{S^0} \rightsquigarrow I^1_{S^1} \rightsquigarrow I^2_{S^2} \rightsquigarrow \cdots \rightsquigarrow I^k_{S^k}$$

with $k \geq 2n-4$. □

We have shown that there is a bound on the length of $\rightsquigarrow$-sequences in congruence lattices of algebras in $\mathcal{V}$. Our next goal is to determine how to recognize $\rightsquigarrow$-related intervals lattice-theoretically.

LEMMA 7.10. *Assume that* $\mathbf{A} \in \mathcal{V}$, *and that* $I_S = I[\mu,\nu]_S$ *and* $J_T = I[\delta,\theta]_T$ *are separated solvability obstructions in* $\mathbf{Con}(\mathbf{A})$. *Choose and fix congruences* δ' *and* μ' *satisfying* $T \cap \delta \subseteq \delta' \subseteq \mathrm{Cg}^{\mathbf{A}}(T)$ *and* $S \cap \mu \subseteq \mu' \subseteq \mathrm{Cg}^{\mathbf{A}}(S)$. *If the conditions*

(d) $\mathrm{Cg}^{\mathbf{A}}(T) \cap \mathrm{Cg}^{\mathbf{A}}(S) \subseteq \delta' \vee \mu'$,
(e) $T \cap (\delta' \vee \mu') \subseteq \delta$, *and*
(f) $S \cap (\delta' \vee \mu') \subseteq \mu$

are satisfied, then the conditions

(b) $T \cap \big(S \circ (T \cap \delta) \circ (S \cap \mu)\big) \subseteq \delta$, *and*
(c) $S \cap \big((T \cap \delta) \circ (S \cap \mu) \circ (T \cap \delta)\big) \subseteq \mu$

from Definition 7.4 are also satisfied.

PROOF. Condition (c) follows immediately from Condition (f), because $T \cap \delta \subseteq \delta'$ and $S \cap \mu \subseteq \mu'$.

We will prove that (b) follows from (d) and (e) by showing that $T \cap \big(S \circ (T \cap \delta) \circ (S \cap \mu)\big) \subseteq T \cap (\delta' \vee \mu')$. Choose

$$(c,d) \in T \cap \big(S \circ (T \cap \delta) \circ (S \cap \mu)\big).$$

There exist a and b such that $c \equiv_S a \equiv_{T\cap\delta} b \equiv_{S\cap\mu} d$, as shown in Figure 7.6. The pair (c,b) belongs to $S \circ (T \cap \delta) \subseteq \mathrm{Cg}^{\mathbf{A}}(S) \circ \delta'$ and also to $T \circ (S \cap \mu) \subseteq \mathrm{Cg}^{\mathbf{A}}(T) \circ \mu'$. Using Theorem 5.28 (2) (to go from the first line below to the second) and Condition (d) of this theorem we get that

$$\begin{aligned}(c,b) &\in \big(\mathrm{Cg}^{\mathbf{A}}(T) \circ \mu'\big) \cap \big(\mathrm{Cg}^{\mathbf{A}}(S) \circ \delta'\big)\\ &\subseteq \big((\mathrm{Cg}^{\mathbf{A}}(T) \vee \delta') \wedge (\mathrm{Cg}^{\mathbf{A}}(S) \vee \mu')\big) \vee \delta' \vee \mu'\\ &\leq \big(\mathrm{Cg}^{\mathbf{A}}(T) \wedge \mathrm{Cg}^{\mathbf{A}}(S)\big) \vee \delta' \vee \mu'\\ &\leq \delta' \vee \mu'.\end{aligned}$$

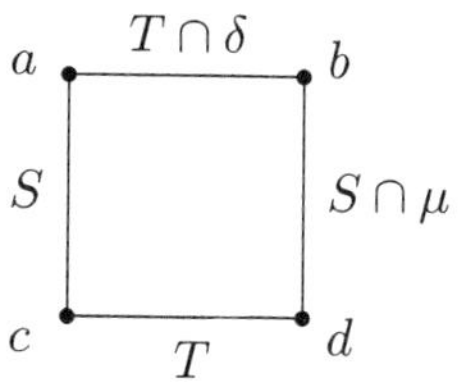

FIGURE 7.6

(In this argument we used the facts that $\mathrm{Cg}^{\mathbf{A}}(T) \supseteq \delta'$ and $\mathrm{Cg}^{\mathbf{A}}(S) \supseteq \mu'$ to replace $(\mathrm{Cg}^{\mathbf{A}}(T) \vee \delta') \wedge (\mathrm{Cg}^{\mathbf{A}}(S) \vee \mu')$ by $\mathrm{Cg}^{\mathbf{A}}(T) \wedge \mathrm{Cg}^{\mathbf{A}}(S)$.) Hence $c \equiv_{\delta' \vee \mu'} b \equiv_{\mu'} d$, so $(c, d) \in T \cap (\delta' \vee \mu')$. Since (c, d) was chosen arbitrarily in $T \cap (S \circ (T \cap \delta) \circ (S \cap \mu))$, we get that

$$T \cap (S \circ (T \cap \delta) \circ (S \cap \mu)) \subseteq T \cap (\delta' \vee \mu').$$

With this and Condition (e) we get Condition (b). This completes the proof. □

The following lemma concerns the behavior of the $\blacktriangleleft$-relation in varieties with a weak difference term, so this lemma could have appeared in Chapter 6. It appears here instead because its primary function is to aid in the proof of the succeeding lemma, which further simplifies the task of recognizing some instances of the $\rightsquigarrow$-relation.

LEMMA 7.11. *Let $\mathcal{U}$ be a variety with a weak difference term and $I[\delta, \theta]_T$ a separated solvability obstruction in* $\mathbf{Con}(\mathbf{A})$, $\mathbf{A} \in \mathcal{U}$.

(1) *$(\delta : T)$ is the largest congruence $\xi \in \mathbf{Con}(\mathbf{A})$ such that $\xi \geq \delta$ and $\xi \cap T \subseteq \delta$.*
(2) *$(\delta : T)$ is a radical congruence.*
(3) *If ρ is a radical congruence and ν is a congruence such that $\nu \not\leq \rho$, then $\nu \wedge \rho \blacktriangleleft \nu$.*

PROOF. Recall that $(\delta : T)$ is defined to be the largest congruence ξ such that $\mathbf{C}(\xi, T; \delta)$ holds, so in particular $\mathbf{C}((\delta : T), T; \delta)$ holds. If $T' := (\delta : T) \cap T$, then the monotonicity of the centralizer relation guarantees that $\mathbf{C}(T', T'; \delta)$ holds. If $T'' := \delta \circ T' \circ \delta$ is the δ-closure of T', then Lemma 6.7 proves that T'' is a congruence containing δ for which $\mathbf{C}(T'', T''; \delta)$ holds. Therefore $\delta \lhd T'' \subseteq \mathrm{Cg}^{\mathbf{A}}(T \cup \delta) \subseteq \theta$, which (since $I[\delta, \theta]$ is a solvability obstruction) forces $T'' = \delta$. Hence $(\delta : T) \cap T = T' \subseteq T'' = \delta$, proving $(\delta : T)$ is one of the congruences above δ whose intersection with T is contained in δ.

Now suppose that ξ is any congruence above δ whose intersection with T is contained in δ. These properties of ξ may be used to show that

$$T \cap (\xi \circ (T \cap \delta) \circ \xi) \subseteq T \cap \xi \subseteq \delta.$$

By Theorem 2.19 (7) it follows that $\mathbf{C}(\xi, T; \delta)$, that is, $\xi \leq (\delta : T)$. This proves item (1).

For item (2) assume instead that $(\delta : T) \lhd \alpha \leq 1$ for some congruence α. Let $T' = \alpha \cap T$. Since $\mathbf{C}(\alpha, \alpha; (\delta : T))$ holds, we also have $\mathbf{C}(T', T'; (\delta : T))$ by the monotonicity of the centralizer. We will derive that $\mathbf{C}(T', T'; \delta)$ holds using Theorem 2.19 (4). To do this we need to prove that $T' \cap (\delta : T) = T' \cap \delta$. The inclusion $\supseteq$ follows from $(\delta : T) \supseteq \delta$, while $\subseteq$ follows from $T' \cap (\delta : T) \subseteq T \cap$

$(\delta : T) \subseteq \delta$. Hence $\mathbf{C}(T', T'; \delta)$ holds indeed. Now, using an argument like the one in the first paragraph of this proof, this contradicts the fact that $I[\delta, \theta]$ is a solvability obstruction unless $T' \subseteq \delta$. But then α is a congruence above δ for which $\alpha \cap T = T' \subseteq \delta$, so by item (1) we get that $\alpha \leq (\delta : T)$. This proves that $(\delta : T) \blacktriangleleft 1$, and therefore that $(\delta : T)$ is a radical congruence.

To prove item (3), if $\nu \not\leq \rho$, then we certainly have $(\nu \wedge \rho) < \nu$. Given any λ satisfying $(\nu \wedge \rho) \lhd \lambda \leq \nu$ we may join with ρ to obtain $\rho \lhd (\rho \vee \lambda) \leq (\rho \vee \nu)$, according to Lemma 6.10 (1). Since $\rho \blacktriangleleft 1$, it follows that $\lambda \leq \rho$. Since $\lambda \leq \nu$ as well, we must have $\lambda \leq (\nu \wedge \rho)$. This proves that $(\nu \wedge \rho) \blacktriangleleft \nu$. □

Because of Lemma 7.11 (1), we may refer to $(\delta : T)$ as the **pseudocomplement of T over δ** when $I[\delta, \theta]_T$ is a separated solvability obstruction.

The following set of criteria for producing $\rightsquigarrow$-related intervals is simpler than the definition of $\rightsquigarrow$ or the criteria of Lemma 7.10, but is still strong enough for all of our intended applications.

LEMMA 7.12. *Suppose that $I[\mu, \nu]_S$ and $I[\delta, \theta]_T$ are separated solvability obstructions in* $\mathbf{Con}(\mathbf{A})$. *If*

(d)′ $\mathrm{Cg}^{\mathbf{A}}(S) \cap \mathrm{Cg}^{\mathbf{A}}(T) \subseteq \big(\mathrm{Cg}^{\mathbf{A}}(S) \wedge (\delta : T)\big) \vee \big(\mathrm{Cg}^{\mathbf{A}}(T) \wedge (\delta : T)\big)$ *and*
(g) $S \cap \mu = S \cap (\delta : T)$,

then $I[\mu, \nu]_S \rightsquigarrow I[\delta, \theta]_T$.

In particular, assume that $I[\delta, \theta]_T$ is any separated solvability obstruction, $\nu \not\leq (\delta : T)$, and $\mu := \nu \wedge (\delta : T)$. Then $I[\mu, \nu]_\nu$ is a separated solvability obstruction, and if $\nu \cap \mathrm{Cg}^{\mathbf{A}}(T) \subseteq \mu$ then $I[\mu, \nu]_\nu \rightsquigarrow I[\delta, \theta]_T$.

PROOF. Assume that (d)′ and (g) hold. We will argue that Condition (a) of Definition 7.4 holds, and Conditions (d), (e) and (f) of Lemma 7.10 all hold in the case where $\delta' = \mathrm{Cg}^{\mathbf{A}}(T) \cap (\delta : T)$ and $\mu' = \mathrm{Cg}^{\mathbf{A}}(S) \cap (\delta : T)$. This choice for δ' satisfies the required condition $T \cap \delta \subseteq \delta' \subseteq \mathrm{Cg}^{\mathbf{A}}(T)$ because $T \subseteq \mathrm{Cg}^{\mathbf{A}}(T)$ and $\delta \subseteq (\delta : T)$. The choice for μ' satisfies the condition $S \cap \mu \subseteq \mu' \subseteq \mathrm{Cg}^{\mathbf{A}}(S)$ because $S \subseteq \mathrm{Cg}^{\mathbf{A}}(S)$, while $S \cap \mu \subseteq (\delta : T)$ is part of assumption (g).

From Condition (g) we get $S \cap (\delta : T) = S \cap \mu \subseteq \mu$ as Condition (a) requires.

Condition (d)′ is simply Condition (d) of Lemma 7.10 restricted to the case where $\delta' = \mathrm{Cg}^{\mathbf{A}}(T) \cap (\delta : T)$ and $\mu' = \mathrm{Cg}^{\mathbf{A}}(S) \cap (\delta : T)$.

Condition (e) asserts that $T \cap (\delta' \vee \mu') \subseteq \delta$. For this, observe that $\delta' \vee \mu' \leq (\delta : T)$ from our choice of primed congruences, so from Lemma 7.11 (1) we get that $T \cap (\delta' \vee \mu') \subseteq T \cap (\delta : T) \subseteq \delta$.

Condition (f) asserts that $S \cap (\delta' \vee \mu') \subseteq \mu$. According to the previous paragraph we have $\delta' \vee \mu' \leq (\delta : T)$. Together with (g), this yields $S \cap (\delta' \vee \mu') \subseteq S \cap (\delta : T) = S \cap \mu \subseteq \mu$.

For the second paragraph of the lemma statement, assume that $I[\delta, \theta]_T$ is any separated solvability obstruction. If $\nu \not\leq (\delta : T)$ and $\mu := \nu \wedge (\delta : T)$, then Lemma 7.11 (2) and (3) prove that $I[\mu, \nu]$ is a solvability obstruction. It is certainly separated by $S := \nu$. To see that (d)′ holds under the assumption $\nu \cap \mathrm{Cg}^{\mathbf{A}}(T) \subseteq \mu$, observe that since $S = \nu$ we have

$$\begin{aligned} \mathrm{Cg}^{\mathbf{A}}(S) \cap \mathrm{Cg}^{\mathbf{A}}(T) &= \nu \cap \mathrm{Cg}^{\mathbf{A}}(T) \\ &\subseteq \mu \qquad \text{(by the assumption)} \\ &:= \nu \cap (\delta : T) \\ &= \mathrm{Cg}^{\mathbf{A}}(S) \cap (\delta : T) \\ &\subseteq \big(\mathrm{Cg}^{\mathbf{A}}(S) \cap (\delta : T)\big) \vee \big(\mathrm{Cg}^{\mathbf{A}}(T) \cap (\delta : T)\big). \end{aligned}$$

For (g), $S \cap \mu = \nu \cap \mu = \mu := \nu \cap (\delta : T) = S \cap (\delta : T)$. □

7.2. Congruence Identities

The first theorem of this section shows that solvability obstructions of minimal rank cannot be contained in a failure of the distributive law. It is easy to combine this with our bound on the length of $\rightsquigarrow$-sequences to derive a nontrivial congruence identity, which we do in Theorem 7.15.

THEOREM 7.13. *Assume that $\mathbf{A} \in \mathcal{V}$ and that $J_T := I[\delta, \theta]_T$ is a separated solvability obstruction in $\mathbf{Con}(\mathbf{A})$. If α, β, and γ are congruences on $\mathbf{A}$ such that*

$$(\alpha \wedge \beta) \vee (\alpha \wedge \gamma) \leq \delta < \theta \leq \alpha \wedge (\beta \vee \gamma),$$

then there is a separated solvability obstruction I_S that is a subinterval of either $I[\alpha \wedge \beta, \beta]$ or $I[\alpha \wedge \gamma, \gamma]$ such that $I_S \rightsquigarrow J_T$.

PROOF. We will apply Lemma 7.12. The pseudocomplement of T over δ cannot lie above both β and γ, for if it did we would have

$$T \subseteq \theta \leq \beta \vee \gamma \leq (\delta : T),$$

which is in conflict with the properties $T \cap (\delta : T) \subseteq \delta$ and $T \not\subseteq \delta$. We may therefore assume that $\gamma \not\leq (\delta : T)$. If $\mu := \gamma \wedge (\delta : T)$, then the second paragraph of the statement of Lemma 7.12 proves that $I[\mu, \gamma]$ is a solvability obstruction. Since $\alpha \wedge \gamma \leq \delta \leq (\delta : T)$, we have $\alpha \wedge \gamma \leq \gamma \wedge (\delta : T) = \mu$, so $I[\mu, \nu]$ is contained in $I[\alpha \wedge \gamma, \gamma]$. Using the second paragraph of the statement of Lemma 7.12 again we obtain from

$$\gamma \cap \mathrm{Cg}^{\mathbf{A}}(T) \leq \gamma \cap \big(\alpha \wedge (\beta \vee \gamma)\big) = \alpha \wedge \gamma \leq \mu$$

that $I[\mu, \nu]_\nu \rightsquigarrow I[\delta, \theta]_T$. □

COROLLARY 7.14. *Assume that $\mathbf{A} \in \mathcal{V}$, and that $\mathbf{Con}(\mathbf{A})$ has a sublattice of congruences labeled as in Figure 7.7. If J_T is a separated solvability obstruction in*

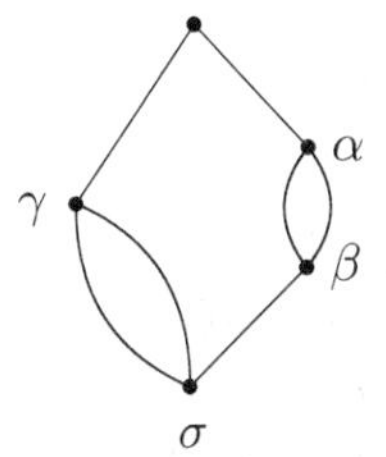

FIGURE 7.7

$I[\beta, \alpha]$, then there is a separated solvability obstruction I_S contained in $I[\sigma, \gamma]$ such that $I_S \rightsquigarrow J_T$.

PROOF. The hypotheses of Theorem 7.13 are satisfied, because $(\alpha \wedge \beta) \vee (\alpha \wedge \gamma) = \beta \vee \sigma = \beta \leq \delta < \theta \leq \alpha = \alpha \wedge (\beta \vee \gamma)$. Therefore there is a separated solvability obstruction I_S contained in either $I[\alpha \wedge \beta, \beta] = I[\beta, \beta]$ or $I[\alpha \wedge \gamma, \gamma] = I[\sigma, \gamma]$ such that $I_S \rightsquigarrow J_T$. The first of these intervals is trivial, so I_S must be a subinterval of the second one. □

THEOREM 7.15. *Let $\mathcal{V}$ be a variety. The following conditions are equivalent.*

(1) *$\mathcal{V}$ satisfies a nontrivial congruence identity.*
(2) *$\mathcal{V}$ satisfies an idempotent Maltsev condition that fails in the variety of semilattices.*
(3) *There exists $\ell < \omega$ such that the lattice $\mathbf{N}_{\ell+5}$, depicted in Figure 7.8, cannot be embedded into $\mathbf{Con}(\mathbf{A})$ for any $\mathbf{A} \in \mathcal{V}$.*

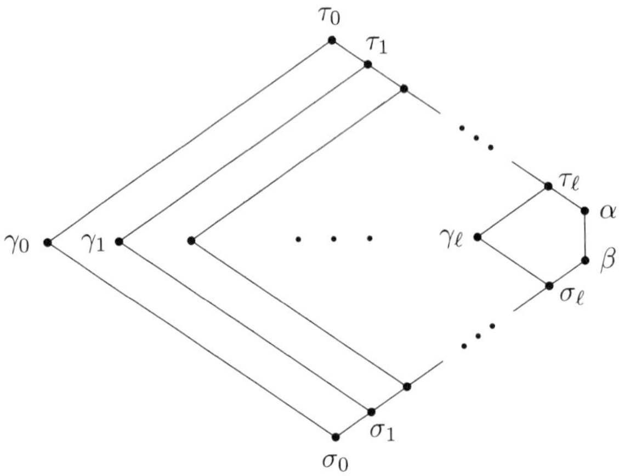

FIGURE 7.8. The lattice $\mathbf{N}_{\ell+5}$

PROOF. If $\mathcal{V}$ satisfies a nontrivial congruence identity ε, then any variety satisfying the same idempotent Maltsev conditions as $\mathcal{V}$ must also satisfy ε, by the Pixley-Wille algorithm. But Ralph Freese and J. B. Nation proved in [**20**] that the variety of semilattices satisfies no nontrivial congruence identity. Therefore $\mathcal{V}$ satisfies an idempotent Maltsev condition that fails in the variety of semilattices.

Assume that (2) holds. Then $\mathcal{V}$ has a Hobby–McKenzie term F of arity n for some $n < \omega$. We shall argue that for $\ell = 2n - 4$ the lattice $\mathbf{N}_{\ell+5}$ cannot be embedded into $\mathbf{Con}(\mathbf{A})$ for any $\mathbf{A} \in \mathcal{V}$. The interval $I[\sigma_\ell, \tau_\ell]$ contains a sublattice $\{\sigma_\ell, \alpha, \beta, \gamma_\ell, \tau_\ell\}$ that is isomorphic to $\mathbf{N}_5$, so this interval is nonmodular. By Theorem 6.16, it follows that $I[\sigma_\ell, \tau_\ell]$ is not ∞-solvable. Therefore, by Theorem 6.14, this interval contains a solvability obstruction $I[\mu_\ell, \nu_\ell]$. Choose a tolerance S_ℓ that separates this interval (e.g., $S_\ell = \nu_\ell$). By Corollary 7.14, applied to the sublattice $\{\sigma_\ell, \sigma_{\ell-1}, \gamma_{\ell-1}, \tau_{\ell-1}, \tau_\ell\}$, there is a separated solvability obstruction $I[\mu_{\ell-1}, \nu_{\ell-1}]_{S_{\ell-1}}$ contained in the interval $I[\sigma_{\ell-1}, \gamma_{\ell-1}]$ such that

$$I[\mu_{\ell-1}, \nu_{\ell-1}]_{S_{\ell-1}} \rightsquigarrow I[\mu_\ell, \nu_\ell]_{S_\ell} .$$

Similarly, for each $i = 1, 2, \ldots, \ell$ successively we can apply Corollary 7.14 to the sublattice $\{\sigma_{\ell-i+1}, \sigma_{\ell-i}, \gamma_{\ell-i}, \tau_{\ell-i}, \tau_{\ell-i+1}\}$ to obtain a separated solvability obstruction

$I[\mu_{\ell-i}, \nu_{\ell-i}]_{S_{\ell-i}}$ contained in the interval $I[\sigma_{\ell-i}, \gamma_{\ell-i}]$ such that

$$I[\mu_{\ell-i}, \nu_{\ell-i}]_{S_{\ell-i}} \rightsquigarrow I[\mu_{\ell-i+1}, \nu_{\ell-i+1}]_{S_{\ell-i+1}} .$$

When we reach $i = \ell$ we will have a sequence

$$I[\mu_0, \nu_0]_{S_0} \rightsquigarrow I[\mu_1, \nu_1]_{S_1} \rightsquigarrow I[\mu_2, \nu_2]_{S_2} \rightsquigarrow \cdots \rightsquigarrow I[\mu_\ell, \nu_\ell]_{S_\ell} .$$

According to Theorem 7.9, there is no such sequence if $\ell \geq 2n - 4$. This completes the proof of (2) $\implies$ (3).

To prove that (3) $\implies$ (1), we observe that:

(i) Each $\mathbf{N}_{\ell+5}$ is subdirectly irreducible.
(ii) Each $\mathbf{N}_{\ell+5}$ is projective in the variety of lattices. (See [**65**].)

By Theorem 2.7, there is an identity $\varepsilon_{\ell+5}$ that holds in a lattice $\mathbf{L}$ if and only if $\mathbf{L}$ has no sublattice isomorphic to $\mathbf{N}_{\ell+5}$. Since (3) asserts that $\mathbf{N}_{\ell+5}$ is not embeddable in $\mathbf{Con}(\mathbf{A})$ for any $\mathbf{A} \in \mathcal{V}$, this identity $\varepsilon_{\ell+5}$ is a congruence identity of $\mathcal{V}$. Since $\varepsilon_{\ell+5}$ fails in $\mathbf{N}_{\ell+5}$, it is a nontrivial congruence identity. □

COROLLARY 7.16 (Cf. [**58**]). *Any congruence n-permutable variety satisfies a nontrivial congruence identity.*

PROOF. It suffices to observe that congruence n-permutability can be characterized by an idempotent Maltsev condition, for example by the Maltsev condition of J. Hagemann and A. Mitschke from [**29**], and that the variety of semilattices is not congruence n-permutable for any n. □

COROLLARY 7.17. *Every congruence join semidistributive variety satisfies a nontrivial congruence identity.*

PROOF. Congruence join semidistributivity can be characterized by a set of idempotent Maltsev conditions, as is shown in [**3**], and the variety of semilattices is not congruence join semidistributive. □

See Theorem 8.14 (7) for explicit congruence identities for congruence join semidistributive varieties.

CHAPTER 8

Congruence Meet and Join Semidistributivity

In this chapter we will characterize the satisfaction of a nontrivial congruence identity in several new ways (Theorems 8.3, 8.10, 8.11, 8.12, and 8.13). Our goal is to understand congruence join semidistributivity and its relationship to the satisfaction of a nontrivial congruence identity. Although most of the results of this chapter concern varieties satisfying congruence identities, we start the chapter by recalling known results about congruence meet semidistributivity and we end the chapter with new results (Theorem 8.14) that provide several characterizations of congruence join semidistributivity for varieties. These results generalize Theorems 9.8, 9.10 and 9.11 of [**34**].

As a byproduct, we solve an old problem of B. Jónsson (Problem 2.18 of [**40**]) by showing that the class of congruence join semidistributive varieties is definable by a Maltsev condition.

It should be noted that any congruence join semidistributive variety is also congruence meet semidistributive (see Exercise 7.14 (10) of [**34**] or (1) $\Longrightarrow$ (6) of Theorem 8.14 of this chapter). Therefore *congruence join semidistributivity* and *congruence semidistributivity* are equivalent properties for varieties.

8.1. Congruence Meet Semidistributivity

Using straightforward arguments, it follows from a combination of the statements in Theorem 2.19 that if α and β are congruences on an algebra $\mathbf{A}$, then $[\alpha,\beta] \leq \alpha \wedge \beta$ and $(\alpha \wedge \beta)/[\alpha,\beta]$ is an abelian congruence on $\mathbf{A}/[\alpha,\beta]$. Therefore, if $\mathcal{V}$ is a variety that omits abelian congruences, it follows that $[\alpha,\beta] = \alpha \wedge \beta$ for any $\alpha, \beta \in \mathrm{Con}(\mathbf{A})$ for any $\mathbf{A} \in \mathcal{V}$. Theorem 2.19 (5) implies that the commutator operation is semidistributive in its first variable i.e.,

$$[\alpha,\gamma] = \delta \;\&\; [\beta,\gamma] = \delta \;\rightarrow\; [\alpha \vee \beta, \gamma] = \delta\,.$$

Hence if $\mathcal{V}$ omits abelian congruences, then the congruence lattices of members of $\mathcal{V}$ must satisfy the ordinary meet semidistributive law:

$$\alpha \wedge \gamma = \delta \;\&\; \beta \wedge \gamma = \delta \;\rightarrow\; (\alpha \vee \beta) \wedge \gamma = \delta\,.$$

The converse of this statement is true, but is much more difficult to prove. It was established for locally finite varieties in Theorem 9.10 of [**34**] using tame congruence theory, and in full generality in Corollary 4.7 of [**53**] using combinatorial arguments. This result links many different properties, as we will show in the next theorem.

Recall the type of meet continuous lattice words defined before Corollary 2.25: if x, y, and z are variables, then let $y_0 = y$, $z_0 = z$, $y_{n+1} = y \vee (x \wedge z_n)$, and $z_{n+1} = z \vee (x \wedge y_n)$. Define $y_\omega = \bigvee_{n<\omega} y_n$ and $z_\omega = \bigvee_{n<\omega} z_n$. When it is necessary to indicate the variables involved and their order we will write $y_\omega(x,y,z)$ for y_ω and $z_\omega(x,y,z)$ for z_ω.

THEOREM 8.1. *Let $\mathcal{V}$ be a variety. The following conditions are equivalent.*

(1) *$\mathcal{V}$ is congruence meet semidistributive.*
(2) *There exists a positive integer n such that if $\mathbf{A} \in \mathcal{V}$ and $\alpha, \beta, \gamma \in \mathrm{Con}(\mathbf{A})$, then*
$$\alpha \cap (\beta \circ \gamma) \subseteq \beta_n ,$$
where $\beta_n := y_n(\alpha, \beta, \gamma)$.
(3) *$[\alpha, \beta] = \alpha \wedge \beta$ for all $\alpha, \beta \in \mathrm{Con}(\mathbf{A})$ and all $\mathbf{A} \in \mathcal{V}$.*
(4) *No member of $\mathcal{V}$ has a nontrivial abelian congruence.*
(5) *No member of $\mathcal{V}$ has a nontrivial abelian tolerance.*
(6) *$\mathbf{C}(\alpha, \beta; \delta) \iff \beta \wedge \big(\alpha \vee (\beta \wedge \delta)\big) \leq \delta$ for all $\mathbf{A} \in \mathcal{V}$ and all $\alpha, \beta, \delta \in \mathrm{Con}(\mathbf{A})$.*
(7) *$\mathcal{V}$ satisfies the meet continuous congruence identity*
$$x \wedge (y \vee z) \approx x \wedge y_\omega ,$$
where $y_\omega = y_\omega(x, y, z)$.
(8) *The meet continuous congruence variety of $\mathcal{V}$ is meet semidistributive.*
(9) *$\mathbf{M}_3$ is not embeddable in $\mathbf{Con}(\mathbf{A})$ for any $\mathbf{A} \in \mathcal{V}$.*
(10) *$\mathcal{V}$ satisfies a family of idempotent Maltsev conditions that, considered together, fail in any nontrivial variety of modules. (Equivalently, $\mathcal{V}$ satisfies* a single *idempotent Maltsev condition that fails in any nontrivial variety of modules.)*

PROOF. The equivalence of items (1), (2) and (3) is Corollary 4.7 of [**53**].

If (3) holds, and γ is an abelian congruence of an algebra $\mathbf{A} \in \mathcal{V}$, then $0 = [\gamma, \gamma] = \gamma \wedge \gamma = \gamma$. Hence γ is trivial, and (4) holds. Conversely, assume that (4) holds. It follows from Theorem 2.19 (8) that $\mathbf{C}(\alpha, \beta; \alpha \wedge \beta)$ holds, so $[\alpha, \beta] \leq \alpha \wedge \beta$ holds for any pair of congruences on any algebra $\mathbf{A}$. Moreover $\mathbf{C}(\alpha, \beta; [\alpha, \beta])$ holds, according to the definition of the commutator, and
$$\mathbf{C}(\alpha, \beta; [\alpha, \beta]) \implies \mathbf{C}(\alpha \wedge \beta, \alpha \wedge \beta; [\alpha, \beta])$$
by Theorem 2.19 (1), so $(\alpha \wedge \beta)/[\alpha, \beta]$ is an abelian congruence on $\mathbf{A}/[\alpha, \beta]$ by Theorem 2.19 (10). Therefore, from (4) we derive (3).

Assume that (4) holds. Since $\mathcal{V}$ omits nontrivial abelian congruences, it omits nontrivial strongly abelian congruences. Therefore, by Theorem 3.13, $\mathcal{V}$ satisfies a nontrivial idempotent Maltsev condition. In this situation, Theorem 3.24 guarantees that an abelian tolerance generates an abelian congruence. Since $\mathcal{V}$ omits nontrivial abelian congruences, $\mathcal{V}$ omits nontrivial abelian tolerances. This shows that (4) $\implies$ (5). The reverse implication is trivial.

To connect item (6) with the preceding conditions, assume that (3) holds and that

(a) $\mathbf{C}(\alpha, \beta; \delta)$ holds.

By Theorem 2.19 (8),

(b) $\mathbf{C}(\alpha, \beta; \beta)$ holds,

so by Theorem 2.19 (6) we derive from (a) and (b) that

(c) $\mathbf{C}(\alpha, \beta; \beta \wedge \delta)$ holds.

Again by Theorem 2.19 (8) we get that

(d) $\mathbf{C}(\beta \wedge \delta, \beta; \beta \wedge \delta)$ holds,

so by Theorem 2.19 (5) we get from (c) and (d) that

(e) $\mathbf{C}(\alpha \vee (\beta \wedge \delta), \beta; \beta \wedge \delta)$ holds.

From this and the definition of the commutator we get that

(f) $\big[(\alpha \vee (\beta \wedge \delta)), \beta\big] \leq \beta \wedge \delta$ holds.

Now from (f) and item (3) we derive that

$$\beta \wedge \big(\alpha \vee (\beta \wedge \delta)\big) = \big[(\alpha \vee (\beta \wedge \delta)), \beta\big] \leq \beta \wedge \delta \leq \delta\,.$$

Conversely, $\beta \wedge \big(\alpha \vee (\beta \wedge \delta)\big) \leq \delta$ implies $\mathbf{C}(\alpha, \beta; \delta)$ for any three congruences on any algebra, as Theorem 2.19 (8) proves. This shows that (3) implies (6).

On the other hand, to see that (6) $\Longrightarrow$ (4), take $\alpha = \beta = \gamma$ and $\delta = 0$ in (6). The result is the assertion that γ is abelian if and only if $\gamma = 0$.

The equivalence of (1), (7) and (8) follows from Theorem 2.23 and the fact that congruence lattices are meet continuous.

Item (8) implies item (9), since $\mathbf{M}_3$ is not meet semidistributive.

Assume that (9) holds. Then, since $\mathbf{M}_3$ satisfies Whitman's condition (W), it follows from Theorem 4.18 that $\mathcal{V}$ satisfies a nontrivial idempotent Maltsev condition. In this situation, a result from [**53**] explains how to construct from a nontrivial abelian congruence on an algebra in $\mathcal{V}$ a copy of $\mathbf{M}_3$ in the congruence lattice of some other algebra in $\mathcal{V}$. Namely, suppose that α is a nonzero abelian congruence on some $\mathbf{A} \in \mathcal{V}$, that $\mathbf{A} \times_\alpha \mathbf{A}$ is the subalgebra of $\mathbf{A}^2$ whose universe is α, that $\eta_1, \eta_2 \in \mathrm{Con}(\mathbf{A} \times_\alpha \mathbf{A})$ are the coordinate projection kernels, and that Δ is the congruence on $\mathbf{A} \times_\alpha \mathbf{A}$ generated by the pairs $\langle (a,a), (b,b)\rangle$ with $(a,b) \in \alpha$. It is a consequence of Theorem 3.5 of [**53**] that η_1, η_2, and Δ generate a sublattice of $\mathbf{Con}(\mathbf{A} \times_\alpha \mathbf{A})$ isomorphic to $\mathbf{M}_3$. Thus, if (9) holds, then (4) must hold.

The Pixley–Wille algorithm shows that the class of varieties satisfying a congruence condition like the one in item (2) can be defined by a single idempotent Maltsev condition. Since (2) is equivalent to (4), this Maltsev condition cannot be satisfied by any variety with nontrivial abelian algebras, hence by any nontrivial variety of modules. Thus (7) implies the stronger of the two statements in (10).

Finally, we show that the weaker of the two statements in (10) implies (4). Assume instead that $\mathcal{V}$ satisfies a family $\mathcal{F}$ of idempotent Maltsev conditions that fail in every nontrivial variety of modules, but that some $\mathbf{A} \in \mathcal{V}$ has a nontrivial abelian congruence α. We may assume, without loss of generality, that $\mathcal{V}$ is an idempotent variety. For if we replace $\mathcal{V}$ and $\mathbf{A}$ by their idempotent reducts, then we do not affect the idempotent Maltsev conditions satisfied by $\mathcal{V}$, nor that α is an abelian congruence on $\mathbf{A}$. Thus, we make this assumption. We may (and do) further assume that $\mathbf{A}$ itself is abelian, since if it is not we may simply replace it by one of its subalgebras that is supported by nontrivial α-block.

At least one of the Maltsev conditions in $\mathcal{F}$ is nontrivial, since these conditions taken together fail in every nontrivial variety of modules. Corollary 4.5 of [**53**] proves that an abelian algebra $\mathbf{A}$ in a variety satisfying some nontrivial idempotent Maltsev condition is **quasi-affine**, which by definition means that $\mathbf{A}$ is a subalgebra of a reduct of an algebra that has the same universe and polynomial operations as some module. Since $\mathbf{A}$ is both idempotent and quasi-affine, it is a subalgebra of an algebra $\mathbf{M}^*$ that is a reduct of some module $\mathbf{M}$ over some ring $\mathbf{R}$. There is no harm in renaming the zero element of $\mathbf{M}$ so that it lies in A. If p is the module term $x - y + z$, then $p\colon \mathbf{M}^3 \to \mathbf{M}$ is a homomorphism, and therefore $p\colon \mathbf{A}^3 \to \mathbf{M}^*$ is also a homomorphism. The image $\mathbf{A}' := p(\mathbf{A}^3)$ contains A, since $p(a,a,a) = a$. Moreover, $\mathbf{A}'$ is a quotient of $\mathbf{A}^3 \in \mathcal{V}$, hence $\mathbf{A}' \in \mathcal{V}$. This produces a possibly larger subalgebra $\mathbf{A} \leq \mathbf{A}' \leq \mathbf{M}^*$. Iterating this, $\mathbf{A} \leq \mathbf{A}' \leq \mathbf{A}'' \leq \cdots$, and taking a union yields an algebra $\mathbf{B} \in \mathcal{V}$ that is a subalgebra of $\mathbf{M}^*$ and is closed under p. Since $0 \in A \subseteq B$, the set B is closed under the abelian group operations $x+y = p(x,0,y)$, $-x = p(0,x,0)$, and 0. Thus $\widehat{\mathbf{B}} := \langle \mathbf{B}; p, 0\rangle$ is an additive subgroup of $\mathbf{M}$ that is also a subalgebra of a reduct of $\mathbf{M}$. This implies that $\widehat{\mathbf{B}}$ is term equivalent to a module over the subring of $\mathbf{R}$ consisting of all elements $r \in R$ such that $rB \subseteq B$. Clearly $\widehat{\mathbf{B}}$ is nontrivial, since $\mathbf{B}$ contains $\mathbf{A}$ as a subalgebra. Moreover, $\widehat{\mathbf{B}}$ satisfies the conditions in $\mathcal{F}$, since its reduct $\mathbf{B}$ does. Therefore the variety generated by $\widehat{\mathbf{B}}$ is (up to term equivalence) a nontrivial variety of modules that satisfies the conditions in $\mathcal{F}$, contrary to (10). This contradicts our assumption that (4) fails to hold, so we are done. □

8.2. More on Congruence Identities

THEOREM 8.2. *Let $\mathcal{V}$ be a variety with a Hobby–McKenzie term. Assume that $\mathbf{A} \in \mathcal{V}$, and that $\mathbf{Con}(\mathbf{A})$ has a sublattice of congruences labeled as in Figure 8.1. If J_T is a separated solvability obstruction in $I[\beta,\alpha]$, then there is a separated*

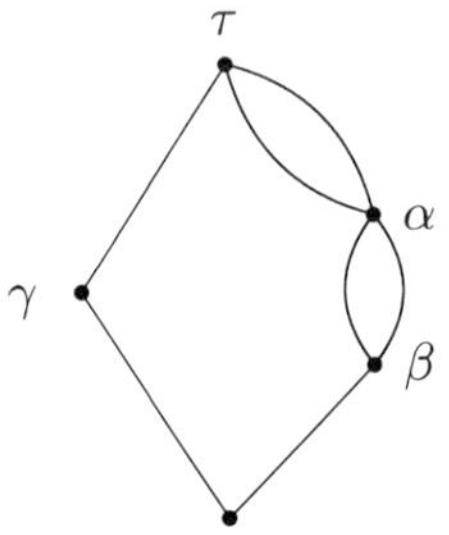

FIGURE 8.1

solvability obstruction I_S contained in $I[\alpha,\tau]$ and a $\rightsquigarrow$-sequence from I_S to J_T.

PROOF. We will show first that if J_T is a separated solvability obstruction in $I[\beta,\alpha]$, then there is a separated solvability obstruction $I'_{S'}$ contained in either $I[\beta,\alpha]$ or $I[\alpha,\tau]$ and a $\rightsquigarrow$-sequence of length 1 or 2 from $I'_{S'}$ to J_T. If it happens that $I'_{S'}$ is contained in $I[\alpha,\tau]$, then we are done. If instead $I'_{S'}$ is contained in $I[\beta,\alpha]$, then we can repeat the argument to find $I''_{S''}$ contained in either $I[\beta,\alpha]$ or $I[\alpha,\tau]$ and a $\rightsquigarrow$-sequence from $I''_{S''}$ to $I'_{S'}$. This argument cannot be repeated indefinitely since concatenating $\rightsquigarrow$-sequences forms longer $\rightsquigarrow$-sequences and there is a bound on the length of $\rightsquigarrow$-sequences. Therefore the argument will produce a separated solvability obstruction in $I[\alpha,\tau]$ that initiates a $\rightsquigarrow$-sequence terminating at J_T.

Recall from the proof of Theorem 7.13, applied to the setting of Corollary 7.14, that if $\mu := \gamma \sqcap (\delta : T)$, then $I[\mu,\gamma]_\gamma \rightsquigarrow I[\delta,\theta]_T$. Since $\mu \leq (\delta : T)$ and $\delta \leq (\delta : T)$, we have

$$\gamma \sqcap (\delta \vee \mu) \leq \gamma \sqcap (\delta : T) = \mu,$$

and therefore $\gamma \wedge \big(\delta \vee (\gamma \wedge \mu)\big) \leq \mu$. By Theorem 2.19 (8) we get that $\mathbf{C}(\delta,\gamma;\mu)$ holds, or equivalently $\delta \leq (\mu : \gamma)$. If $\alpha \not\leq (\mu : \gamma)$, then for $\lambda := \alpha \wedge (\mu : \gamma)$ $(\geq \delta \geq \beta)$ we have that $I[\lambda,\alpha]$ is a nontrivial subinterval of $I[\beta,\alpha]$. According to the second paragraph of Lemma 7.12, the interval $I[\lambda,\alpha]$ is a solvability obstruction. That paragraph further guarantees that, since $\alpha \cap \mathrm{Cg}^{\mathbf{A}}(\gamma) \leq \lambda$, we have $I[\lambda,\alpha]_\alpha \rightsquigarrow I[\mu,\gamma]_\gamma$. Therefore, in the case where $\alpha \not\leq (\mu : \gamma)$ we have located a separated solvability obstruction $I[\lambda,\alpha]_\alpha$ in $I[\beta,\alpha]$ and a $\rightsquigarrow$-sequence

$$I[\lambda,\alpha]_\alpha \rightsquigarrow I[\mu,\gamma]_\gamma \rightsquigarrow I[\delta,\theta]_T = J_T$$

of length two.

Now consider the case where $\alpha \leq (\mu : \gamma)$. Since $\gamma \not\leq (\mu : \gamma)$ and $\gamma \leq \tau$ we have $\tau \not\leq (\mu : \gamma)$, so $\kappa := \tau \cap (\mu : \gamma)$ satisfies $\alpha \leq \kappa < \tau$ in this case. According to the second paragraph of Lemma 7.12, the interval $I[\kappa,\tau]$ is a solvability obstruction. Since $\gamma \not\leq \kappa$, $I[\kappa,\tau]_\gamma$ is a separated solvability obstruction. We prove that $I[\kappa,\tau]_\gamma \rightsquigarrow I[\delta,\tau]_T$ by verifying criteria (d)' and (g) from Lemma 7.12. In our setting these statements are

(d)' $\gamma \cap \mathrm{Cg}^{\mathbf{A}}(T) \subseteq \big(\gamma \wedge (\delta : T)\big) \vee \big(\mathrm{Cg}^{\mathbf{A}}(T) \wedge (\delta : T)\big)$ and
(g) $\gamma \cap \kappa = \gamma \cap (\delta : T)$.

Item (d)' follows from the observation that

$$\gamma \cap \mathrm{Cg}^{\mathbf{A}}(T) \subseteq \gamma \cap \alpha \subseteq \mu := \gamma \cap (\delta : T)\,,$$

since $\gamma \cap (\delta : T)$ is the first joinand in the right hand side of (d)'. For item (g), observe that

$$\gamma \cap \kappa = \gamma \cap \tau \cap (\mu : \gamma) = \gamma \cap (\mu : \gamma) = \mu = \gamma \cap (\delta : T)\,.$$

This proves that $I[\kappa,\tau]_\gamma \rightsquigarrow I[\delta,\tau]_T$. □

Let **L** be a lattice. A **herringbone** in **L** is the union of three descending chains $\{\alpha^i\} \cup \{\beta^{2i}\} \cup \{\gamma^{2i+1}\}$ in **L** where $\{\alpha^i\} \cup \{\beta^{2i}\}$ and $\{\alpha^i\} \cup \{\gamma^{2i+1}\}$ are sublattices of **L** ordered as in Figure 8.2. In other words, a herringbone is a partial sublattice

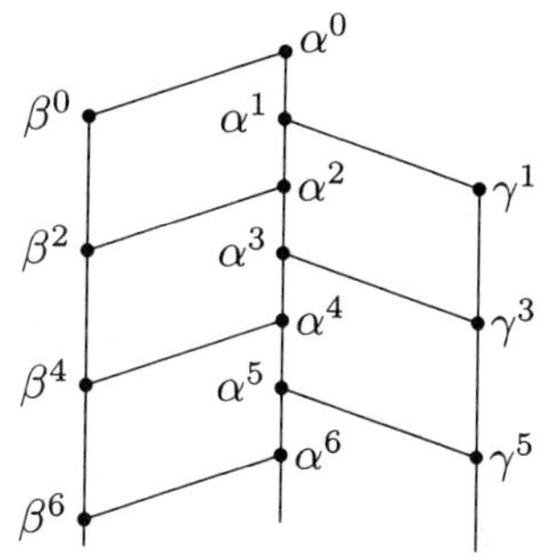

FIGURE 8.2. A herringbone

of **L**, ordered as in Figure 8.2, with

(β) $\alpha^{2i+1} \wedge \beta^{2i} = \beta^{2i+2}$, $\alpha^{2i+2} \vee \beta^{2i} = \alpha^{2i}$, and
(γ) $\alpha^{2i+2} \wedge \gamma^{2i+1} = \gamma^{2i+3}$, $\alpha^{2i+3} \vee \gamma^{2i+1} = \alpha^{2i+1}$.

The **length** of the herringbone is ∞ if there are infinitely many distinct α's. Otherwise, the **length** of the herringbone is the supremum of the superscripts k such that $\alpha^0 > \alpha^1 > \cdots > \alpha^k$. (It is easy to see from (β) and (γ) that if some $\alpha^k = \alpha^{k+1}$ then $\alpha^k = \alpha^{k+1} = \alpha^{k+2} = \cdots$.) Our first goal is to prove that if a variety $\mathcal{V}$ satisfies a nontrivial congruence identity, then there is a fixed bound on the length of any herringbone that appears in the congruence lattice of a member of $\mathcal{V}$.

Let x, y and z be lattice variables. Define lattice words by $y^0 = y$, $z^0 = z$, $y^{n+1} = y \wedge (x \vee z^n)$, and $z^{n+1} = z \wedge (x \vee y^n)$. The construction of these words is dual to the words y_n and z_n, which preceded the statement of Theorem 8.1. As there, we will write $y^n(x, y, z)$ and $z^n(x, y, z)$ when it is necessary to indicate the order of the variables.

THEOREM 8.3. *Let $\mathcal{V}$ be a variety. The following conditions are equivalent.*

(1) *$\mathcal{V}$ satisfies a nontrivial congruence identity.*
(2) *There is a positive integer N such that no algebra in $\mathcal{V}$ has a herringbone of length greater than N in its congruence lattice.*
(3) *$\mathcal{V}$ satisfies the congruence identity*
$$z^M \approx z^{M+1}$$
for some M.

PROOF. If $\mathcal{V}$ satisfies a nontrivial congruence identity, then by Theorems 7.15 and 5.25 the variety $\mathcal{V}$ has a Hobby–McKenzie term of arity n for some $n < \omega$. We argue that no algebra in $\mathcal{V}$ has a herringbone of length greater than $2n - 3$ in its congruence lattice. If this is not the case, then some $\mathbf{A} \in \mathcal{V}$ has a herringbone $\{\alpha^i\} \cup \{\beta^{2i}\} \cup \{\gamma^{2i+1}\}$ with $\alpha^0 > \alpha^1 > \cdots > \alpha^{2n-3} > \alpha^{2n-2}$. Figure 8.3 isolates five congruences near the bottom of the herringbone that form a sublattice of $\mathbf{Con}(\mathbf{A})$ isomorphic to $\mathbf{N}_5$. For the congruences in this figure, if $\alpha^{2n-3} \lll \alpha^{2n-4}$, then by

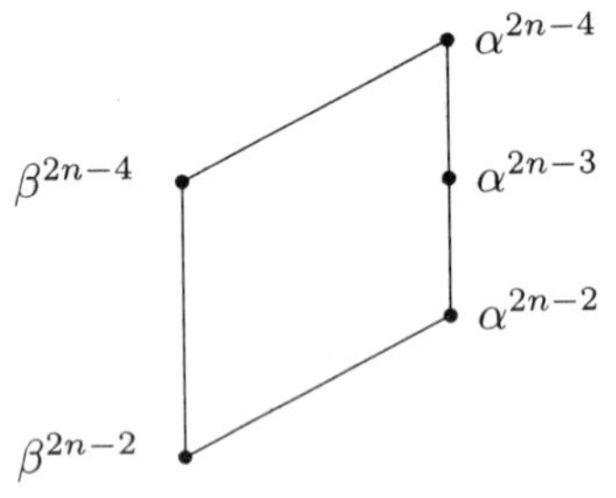

FIGURE 8.3. A piece near the bottom of the herringbone

Lemma 6.10 (2) we derive that
$$\beta^{2n-2} = \beta^{2n-4} \wedge \alpha^{2n-3} \lll \beta^{2n-4} \wedge \alpha^{2n-4} = \beta^{2n-4}.$$

But this was proved to be impossible in Theorem 6.25 (1). Therefore we cannot have $\alpha^{2n-3} \lll \alpha^{2n-4}$. By Theorem 6.14 there is a solvability obstruction $I[\mu_{2n-4}, \nu_{2n-4}]$ contained in the interval $I[\alpha^{2n-3}, \alpha^{2n-4}]$. This obstruction is separated by $S_{2n-4} := \nu_{2n-4}$. We are now in a position to apply Theorem 8.2 to a copy of $\mathbf{N}_5$ that is slightly higher up in this herringbone (Figure 8.4). The separated solvability obstruction $I[\mu_{2n-4}, \nu_{2n-4}]_{S_{2n-4}}$ that is contained in $I[\alpha^{2n-3}, \alpha^{2n-4}]$ plays the role of J_T of that theorem. The theorem guarantees the existence of a separated solvability obstruction $I[\mu_{2n-5}, \nu_{2n-5}]_{S_{2n-5}}$ in $I[\alpha^{2n-4}, \alpha^{2n-5}]$ and a $\rightsquigarrow$-sequence
$$I[\mu_{2n-5}, \nu_{2n-5}]_{S_{2n-5}} \rightsquigarrow \cdots \rightsquigarrow I[\mu_{2n-4}, \nu_{2n-4}]_{S_{2n-4}}.$$

We can work our way up the herringbone, alternately using copies of $\mathbf{N}_5$ on the right and left and using Theorem 8.2 at each step, to derive that for each

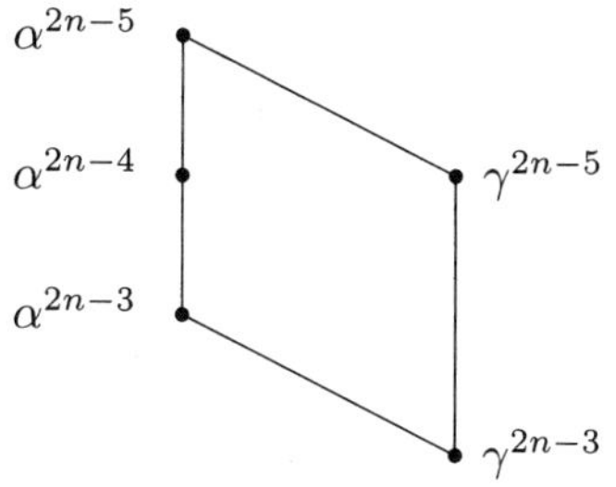

FIGURE 8.4. A piece higher up

$i = 4, 5, \ldots, 2n-1$, if $I[\alpha^{2n-i+1}, \alpha^{2n-i}]$ contains a separated solvability obstruction $I[\mu_{2n-i}, \nu_{2n-i}]_{S_{2n-i}}$, then $I[\alpha^{2n-i}, \alpha^{2n-i-1}]$ contains a separated solvability obstruction $I[\mu_{2n-i-1}, \nu_{2n-i-1}]_{S_{2n-i-1}}$ for which there exists a $\rightsquigarrow$-sequence

$$I[\mu_{2n-i-1}, \nu_{2n-i-1}]_{S_{2n-i-1}} \rightsquigarrow \cdots \rightsquigarrow I[\mu_{2n-i}, \nu_{2n-i}]_{S_{2n-i}}\,.$$

When we reach the top of this herringbone we will have located a $\rightsquigarrow$-sequence

$$I[\mu_0, \nu_0]_{S_0} \rightsquigarrow \cdots \rightsquigarrow I[\mu_1, \nu_1]_{S_1} \rightsquigarrow \cdots\cdots \rightsquigarrow I[\mu_{2n-4}, \nu_{2n-4}]_{S_{2n-4}}$$

of length at least $2n-4$. But we proved in Theorem 7.9 that there is no $\rightsquigarrow$-sequence of this length. This contradiction proves that there is no herringbone of length greater than $2n-3$ in $\mathbf{Con}(\mathbf{A})$.

Now we prove that (2) $\Longrightarrow$ (3). Start with three congruences α, β and γ, and define $\beta^n := y^n(\alpha, \beta, \gamma)$ and $\gamma^n := z^n(\alpha, \beta, \gamma)$ where $y^n(x, y, z)$ and $z^n(x, y, z)$ are the terms defined before the statement of this theorem. Let $\alpha^n := \alpha \vee \beta^n$ if n is even, and $\alpha^n := \alpha \vee \gamma^n$ if n is odd. Since $\beta^0 = \beta \geq \beta \wedge (\alpha \vee \gamma^0) = \beta^1$, and $\gamma^0 \geq \gamma^1$, it is easy to see inductively that

$$\beta^{n+1} = \beta \wedge (\alpha \vee \gamma^n) \geq \beta \wedge (\alpha \vee \gamma^{n+1}) = \beta^{n+2}\,,$$

and $\gamma^{n+1} \geq \gamma^{n+2}$. Thus the β and γ-sequences are descending chains, which forces the α-sequence to be a descending chain. We claim that

$$\{\alpha^n \mid \text{all } n\} \cup \{\beta^n \mid \text{even } n\} \cup \{\gamma^n \mid \text{odd } n\}$$

is a herringbone. To see this, we must verify conditions (β) and (γ) from the definition of a herringbone:

(β) $\alpha^{2i+1} \wedge \beta^{2i} = \beta^{2i+2}$, $\alpha^{2i+2} \vee \beta^{2i} = \alpha^{2i}$,
(γ) $\alpha^{2i+2} \wedge \gamma^{2i+1} = \gamma^{2i+3}$, $\alpha^{2i+3} \vee \gamma^{2i+1} = \alpha^{2i+1}$.

To show that $\alpha^{2i+1} \wedge \beta^{2i} = \beta^{2i+2}$, observe that

$$\beta^{2i+2} = \beta \wedge (\alpha \vee \gamma^{2i+1}) = \beta \wedge \alpha^{2i+1} \leq \alpha^{2i+1}\,,$$

and (as observed earlier) $\beta^{2i+2} \leq \beta^{2i}$. Thus $\beta^{2i+2} \leq \alpha^{2i+1} \wedge \beta^{2i}$. Conversely,

$$\alpha^{2i+1} \wedge \beta^{2i} = (\alpha \vee \gamma^{2i+1}) \wedge \beta^{2i} \leq (\alpha \vee \gamma^{2i+1}) \wedge \beta = \beta^{2i+2}\,.$$

To show that $\alpha^{2i+2} \vee \beta^{2i} = \alpha^{2i}$, observe that

$$\alpha^{2i+2} \vee \beta^{2i} \geq \alpha \vee \beta^{2i} = \alpha^{2i}\,,$$

while $\alpha^{2i+2} \leq \alpha^{2i}$ and $\beta^{2i} \leq \alpha \vee \beta^{2i} = \alpha^{2i}$, so $\alpha^{2i+2} \vee \beta^{2i} \leq \alpha^{2i}$. This establishes ($\beta$), and ($\gamma$) can be established the same way.

It follows that if N bounds the length of any herringbone in any congruence lattice of a member of $\mathcal{V}$, then however we choose our original three congruences α, β and γ, the sequences defined above must satisfy $\alpha^N = \alpha^{N+1} = \alpha^{N+2} = \cdots$, and therefore $\gamma^M = \gamma^{M+2}$ holds for each $M \geq N$. Since the γ's form a descending chain, we conclude that $\gamma^M = \gamma^{M+1}$ holds for each $M \geq N$. Therefore $z^M \approx z^{M+1}$ holds in all congruence lattices of members of $\mathcal{V}$.

Item (1) follows from (3) and the fact that the identity $z^M \approx z^{M+1}$ is nontrivial. To see that it fails in some lattice, take any lattice containing a herringbone of length at least $M+1$. Let $\alpha := \alpha^{M+1}$, $\beta := \beta^0$ and $\gamma := \gamma^1$. It is easy to see that $z^M(\alpha,\beta,\gamma) \neq z^{M+1}(\alpha,\beta,\gamma)$. □

This argument shows that if $\mathcal{V}$ has a Hobby–McKenzie term of arity n, then $\mathcal{V}$ satisfies the congruence identity $z^{2n-3} \approx z^{2n-2}$. These exponents are not optimal, since if $n = 3$ then our proof shows that $\mathcal{V}$ satisfies the congruence identity $z^3 \approx z^4$. But it can be argued (by considering all Hobby–McKenzie terms of arity 3) that when $n = 3$ the variety $\mathcal{V}$ is congruence modular, and consequently satisfies $z^1 \approx z^2$ as a congruence identity.

Recall from Section 2.2 that the $\mathrm{SD}_\vee$-configuration is the pair

$$(\mathbf{P}(\mathrm{SD}_\vee), p \approx j)$$

where $\mathbf{P}(\mathrm{SD}_\vee)$ is the partial lattice given by the presentation $\langle G \mid R\rangle$ where $G = \{p,q,r,j,m\}$ and R consists of the relations $p \vee q = j = p \vee r$ and $q \wedge r = m \leq p$.

LEMMA 8.4. *$\mathbf{P}(\mathrm{SD}_\vee)$ is projective relative to any variety of lattices satisfying an identity of the form $z^M \approx z^{M+1}$.*

In fact, if $\sigma\colon \mathbf{L} \to \mathbf{K}$ is a surjective homomorphism in this variety and $\varphi\colon \mathbf{P}(\mathrm{SD}_\vee) \to \mathbf{K}$ is a partial lattice homomorphism, then for any function $f\colon \{p,q,r,s\} \to \mathbf{L}$ satisfying $\sigma \circ f = \varphi$ on the domain of f there is a homomorphism of partial lattices $\overline{\varphi}\colon \mathbf{P}(\mathrm{SD}_\vee) \to \mathbf{L}$ such that $\sigma \circ \overline{\varphi} = \varphi$ and $\overline{\varphi}(p) \geq f(p)$, $\overline{\varphi}(q) \leq f(q)$, and $\overline{\varphi}(r) \leq f(r)$.

PROOF. Define $\mathbf{P} := \mathbf{P}(\mathrm{SD}_\vee)$, and suppose that $\varphi\colon \mathbf{P} \to \mathbf{K}$ is a homomorphism of partial lattices and $f\colon \{p,q,r,s\} \to \mathbf{L}$ is a function satisfying $\sigma \circ f = \varphi$ on the domain of f. Let $a = f(p)$, $b = f(q)$ and $c = f(r)$. Recursively define $b^0 := b$, $c^0 := c$, $b^{n+1} := b \wedge (a \vee c^n)$, and $c^{n+1} := c \wedge (a \vee b^n)$, so that $b^n = y^n(a,b,c)$ and $c^n := z^n(a,b,c)$ where $y^n(x,y,z)$ and $z^n(x,y,z)$ are the terms defined for the proof of Theorem 8.3. As we showed in the proof of Theorem 8.3 (1) $\Longrightarrow$ (2), $b = b^0 \geq b^1 \geq \cdots$ and $c = c^0 \geq c^1 \geq \cdots$. From the identity $z^M \approx z^{M+1}$ we get that these chains terminate, in which case

$$b^M = b^{M+1} = b \wedge (a \vee c^M) \leq a \vee c^M$$

and similarly $c^M \leq a \vee b^M$. Thus $b^M \leq b$, $c^M \leq c$ and $a \vee b^M = a \vee c^M$. Define $\overline{\varphi}\colon \mathbf{P} \to \mathbf{L}$ by $\overline{\varphi}(p) = a \vee (b^M \wedge c^M)$ $(\geq f(p))$, $\overline{\varphi}(q) = b^M$ $(\leq f(q))$, $\overline{\varphi}(r) = c^M$ $(\leq f(r))$, $\overline{\varphi}(j) = a \vee b^M$, and $\overline{\varphi}(m) = b^M \wedge c^M$. It is clear that

$$\overline{\varphi}(j) = a \vee b^M = \overline{\varphi}(p) \vee \overline{\varphi}(q) = \overline{\varphi}(p) \vee \overline{\varphi}(r)$$

and that

$$\overline{\varphi}(m) = b^M \wedge c^M = \overline{\varphi}(q) \wedge \overline{\varphi}(r) \leq \overline{\varphi}(p)\,.$$

Therefore $\overline{\varphi}\colon \mathbf{P} \to \mathbf{L}$ is a homomorphism of partial lattices. To finish the proof we must show that $\sigma \circ \overline{\varphi} = \varphi$. It suffices to prove that $\sigma \circ \overline{\varphi}(x) = \sigma \circ f(x)$ $(= \varphi(x))$ when $x \in \{p,q,r\}$ is a generator of $\mathbf{P}$.

By simultaneous induction, $\sigma(b^n) = \sigma(b)$ and $\sigma(c^n) = \sigma(c)$ for all n, where one of the induction steps is proved by

$$\begin{aligned}\sigma(b^{n+1}) &= \sigma(b) \wedge \big(\sigma(a) \vee \sigma(c^n)\big) \\ &= \sigma(b) \wedge \big(\sigma(a) \vee \sigma(c)\big) \\ &= \sigma(b) \wedge \big(\sigma(a) \vee \sigma(b)\big) = \sigma(b)\,.\end{aligned}$$

Thus we have $\sigma \circ f(q) = \sigma(b) = \sigma(b^M) = \sigma \circ \overline{\varphi}(q)$, and similarly $\sigma \circ f(r) = \sigma(c) = \sigma(c^M) = \sigma \circ \overline{\varphi}(r)$. Since $q \wedge r \leq p$ and φ is a homomorphism we have $\varphi(q) \wedge \varphi(r) \leq \varphi(p)$. This yields the second equality in

$$\begin{aligned}\sigma \circ f(p) &= \sigma(a) \\ &= \sigma(a) \vee \big(\sigma(b) \wedge \sigma(c)\big) \\ &= \sigma(a) \vee \big(\sigma(b^M) \wedge \sigma(c^M)\big) \\ &= \sigma \circ \overline{\varphi}(p)\,.\end{aligned}$$

The lemma is proved. □

Our first use of Lemma 8.4 will require the following consequence of Theorem 8.3, which seems to be new even for locally finite varieties.

THEOREM 8.5. *Assume that $\mathcal{V}$ satisfies a nontrivial congruence identity. If I is an* $\mathrm{SD}_\vee$*-failure in* $\mathbf{Con}(\mathbf{A})$ *for some* $\mathbf{A} \in \mathcal{V}$*, then I is abelian.*

PROOF. We proved in Corollary 6.33 that if $\mathcal{V}$ has a weak difference term, then ∞-solvable $\mathrm{SD}_\vee$-failures are abelian. Since the assumption that $\mathcal{V}$ satisfies a nontrivial congruence identity implies that $\mathcal{V}$ has a Hobby–McKenzie term, and therefore a weak difference term (Corollary 6.3), to prove the present theorem it will suffice to show that all $\mathrm{SD}_\vee$-failures are ∞-solvable. To set up notation for this, assume that $\mathbf{A} \in \mathcal{V}$ has congruences α, β and γ satisfying $\alpha \vee \beta = \alpha \vee \gamma$ and $\beta \wedge \gamma \leq \alpha$. Our goal is to prove that $\alpha \lll \alpha \vee \beta$. Factoring by $\beta \wedge \gamma$ if necessary, we also assume that $\beta \wedge \gamma = 0$.

Our strategy will be to show that if it is not true that $\alpha \lll \alpha \vee \beta$, then $\mathbf{Con}(\mathbf{A})$ contains an infinite $\rightsquigarrow$-sequence. This will suffice to prove the theorem, since Theorem 7.9 provides a finite bound on the length of $\rightsquigarrow$-sequences. Therefore, assume that α is *not* ∞-solvably related to $\alpha \vee \beta = \alpha \vee \gamma$. Define new congruences as follows:

- $\beta^0 := \beta \wedge \mathrm{rad}(\alpha)$, and $\gamma^0 := \gamma \wedge \mathrm{rad}(\alpha)$.
- $\beta^{k+1} := \beta \wedge (\gamma^k : \gamma)$ and $\gamma^{k+1} := \gamma \wedge (\beta^k : \beta)$.

Our first goal will be to prove that $\beta^k \blacktriangleleft \beta$ and $\gamma^k \blacktriangleleft \gamma$ for all k.

CLAIM 8.6. *$\beta^0 \blacktriangleleft \beta$ and $\gamma^0 \blacktriangleleft \gamma$.*

We have assumed that α is not ∞-solvably related to $\alpha \vee \beta$, so $\alpha \vee \beta \not\leq \mathrm{rad}(\alpha)$. Since $\alpha \leq \mathrm{rad}(\alpha)$, we conclude that $\beta \not\leq \mathrm{rad}(\alpha)$. Since $\mathrm{rad}(\alpha)$ is a radical congruence, Lemma 7.11 (3) guarantees that $\beta^0 := \beta \wedge \mathrm{rad}(\alpha) \blacktriangleleft \beta$ and similarly $\gamma^0 \blacktriangleleft \gamma$.

CLAIM 8.7. *$\alpha \leq (\beta^k : \beta)$ and $\alpha \leq (\gamma^k : \gamma)$ for all k.*

From $\beta^0 := \beta \wedge \mathrm{rad}(\alpha)$ we derive $\mathbf{C}(\mathrm{rad}(\alpha), \beta; \beta^0)$ using Theorem 2.19 (8). Hence $\alpha \leq \mathrm{rad}(\alpha) \leq (\beta^0 : \beta)$. An identical argument proves that $\alpha \leq (\gamma^0 : \gamma)$.

Now assume that $\alpha \leq (\beta^k : \beta)$ for some k. Since $\gamma^{k+1} = \gamma \wedge (\beta^k : \beta)$ we compute that

$$\gamma^{k+1} \leq \gamma \wedge (\alpha \vee \gamma^{k+1}) \leq \gamma \wedge \big((\beta^k : \beta) \vee \gamma^{k+1}\big) = \gamma \wedge (\beta^k : \beta) = \gamma^{k+1}\,,$$

so $\gamma^{k+1} = \gamma \wedge (\alpha \vee \gamma^{k+1})$. By Theorem 2.19 (8) we conclude that $\mathbf{C}(\alpha, \gamma; \gamma^{k+1})$ holds, hence $\alpha \leq (\gamma^{k+1} : \gamma)$. This proof that $\alpha \leq (\beta^k : \beta)$ implies $\alpha \leq (\gamma^{k+1} : \gamma)$ works for every k, and for β and γ interchanged, so the claim follows by induction.

CLAIM 8.8. *If $\beta^k \blacktriangleleft \beta$, then we have $\gamma^{k+1} \blacktriangleleft \gamma$. Similarly, if $\gamma^k \blacktriangleleft \gamma$, then $\beta^{k+1} \blacktriangleleft \beta$.*

If $\beta^k \blacktriangleleft \beta$, then $I[\beta^k, \beta]_\beta$ is a separated solvability obstruction, so according to Lemma 7.11 (2) the pseudocomplement $(\beta^k : \beta)$ is a radical congruence. We proved in Claim 8.7 that $\alpha \leq (\beta^k : \beta)$. If it is also the case that $\gamma \leq (\beta^k : \beta)$, then

$$\beta \leq \alpha \vee \beta = \alpha \vee \gamma \leq (\beta^k : \beta),$$

forcing $\mathbf{C}(\beta, \beta; \beta^k)$, or equivalently $\beta^k \lhd \beta$. This is contrary to $\beta^k \blacktriangleleft \beta$, so it must be that $\gamma \not\leq (\beta^k : \beta)$. Now Lemma 7.11 (3) guarantees that

$$\gamma^{k+1} := \gamma \wedge (\beta^k : \beta) \blacktriangleleft \gamma.$$

The proof that $\gamma^k \blacktriangleleft \gamma$ implies $\beta^{k+1} \blacktriangleleft \beta$ is the same.

CLAIM 8.9. *$I[\gamma^{k+1}, \gamma]_\gamma \rightsquigarrow I[\beta^k, \beta]_\beta$ and $I[\beta^{k+1}, \beta]_\beta \rightsquigarrow I[\gamma^k, \gamma]_\gamma$ for all k.*

The second paragraph of Lemma 7.12 describes a method to produce $\rightsquigarrow$-related intervals, which we employ now. Since $I[\beta^k, \beta]_\beta$ is a separated solvability obstruction and $\gamma \not\leq (\beta^k : \beta)$, then for the congruence $\gamma^{k+1} := \gamma \wedge (\beta^k : \beta)$ we have that $I[\gamma^{k+1}, \gamma]_\gamma$ is also a separated solvability obstruction. Moreover, as is shown in Lemma 7.12, to establish that $I[\gamma^{k+1}, \gamma]_\gamma \rightsquigarrow I[\beta^k, \beta]_\beta$ it is sufficient to verify that $\gamma \cap \mathrm{Cg}^{\mathbf{A}}(\beta) \subseteq \gamma^{k+1}$. But $\gamma \cap \mathrm{Cg}^{\mathbf{A}}(\beta) = \gamma \cap \beta = 0$, so this certainly holds.

Under the assumption that α is not ∞-solvably related to $\alpha \vee \beta$ we have produced an infinite $\rightsquigarrow$-sequence

$$\cdots\cdots \rightsquigarrow I[\beta^4, \beta]_\beta \rightsquigarrow I[\gamma^3, \gamma]_\gamma \rightsquigarrow I[\beta^2, \beta]_\beta \rightsquigarrow I[\gamma^1, \gamma]_\gamma \rightsquigarrow I[\beta^0, \beta]_\beta .$$

This completes the proof of the theorem, as we explained in the second paragraph of this proof. □

Let $\mathcal{V}$ be a variety satisfying a nontrivial congruence identity. Theorem 8.5 shows that if $\mathbf{A} \in \mathcal{V}$, then the $\mathrm{SD}_\vee$-failures in $\mathbf{Con}(\mathbf{A})$ are abelian. Since $\mathcal{V}$ has a weak difference term, these abelian intervals consist of permuting equivalence relations, hence these intervals are modular. It turns out that this property characterizes the class of varieties satisfying nontrivial congruence identities.

THEOREM 8.10. *The following are equivalent for a variety $\mathcal{V}$.*

(1) *$\mathcal{V}$ satisfies a nontrivial congruence identity.*
(2) *For every $\mathbf{A} \in \mathcal{V}$, $\mathbf{Con}(\mathbf{A})$ has an* $\mathrm{SD}_\vee$/Modular *factorization in the category of meet continuous lattices.*
(3) *For every $\mathbf{A} \in \mathcal{V}$, $\mathbf{Con}(\mathbf{A})$ has an* $\mathrm{SD}_\vee$/Modular *factorization in the category of ordinary lattices.*
(4) *For every $\mathbf{A} \in \mathcal{V}$, each $\mathrm{SD}_\vee$-failure in $\mathbf{Con}(\mathbf{A})$ is modular.*
(5) *If $\mathbf{L}$ belongs to $\mathrm{CON}(\mathcal{V})$, then each $\mathrm{SD}_\vee$-failure in $\mathbf{L}$ is modular.*

PROOF. If (1) holds, then from Theorem 7.15 (1) $\Longrightarrow$ (2), Theorem 5.25 (2) $\Longrightarrow$ (3), and Corollary 6.3 it can be deduced that $\mathcal{V}$ has a weak difference term. Therefore the natural homomorphism

$$\nu\colon \mathbf{Con}(\mathbf{A}) \to \mathbf{Con}(\mathbf{A})/\overset{s}{\sim}$$

provides a factorization of $\mathbf{Con}(\mathbf{A})$ of the form $\mathrm{SD}_\wedge$/Modular in the category of meet continuous lattices. We need to argue that the quotient $\mathbf{Con}(\mathbf{A})/\overset{s}{\sim}$ is not only meet semidistributive, but is also join semidistributive.

Each $\mathrm{SD}_\vee$-failure in $\mathbf{Con}(\mathbf{A})/\overset{s}{\sim}$ is of the form $I = I[\varphi(p), \varphi(j)]$ for some homomorphism of partial lattices $\varphi\colon \mathbf{P} \to \mathbf{Con}(\mathbf{A})/\overset{s}{\sim}$ from the partial lattice $\mathbf{P} := \mathbf{P}(\mathrm{SD}_\vee)$ of the $\mathrm{SD}_\vee$-configuration. By the projectivity of $\mathbf{P}$, proved in Lemma 8.4, to each such homomorphism there is a homomorphism $\overline{\varphi}\colon \mathbf{P} \to \mathbf{Con}(\mathbf{A})$ such that $\nu \circ \overline{\varphi} = \varphi$ where $\nu\colon \mathbf{Con}(\mathbf{A}) \to \mathbf{Con}(\mathbf{A})/\overset{s}{\sim}$ is the natural map. The homomorphism $\overline{\varphi}$ identifies an $\mathrm{SD}_\vee$-failure $I[\overline{\varphi}(p), \overline{\varphi}(j)]$ in $\mathbf{Con}(\mathbf{A})$, which by Theorem 8.5 must be abelian. But then $\overline{\varphi}(p) \overset{s}{\sim} \overline{\varphi}(j)$, so since $\overset{s}{\sim}$ is contained in $\ker(\nu)$ we derive that

$$\varphi(p) = \nu \circ \overline{\varphi}(p) = \nu \circ \overline{\varphi}(j) = \varphi(j)\,.$$

This shows that our original $\mathrm{SD}_\vee$-failure $I[\varphi(p), \varphi(j)]$ in $\mathbf{Con}(\mathbf{A})/\overset{s}{\sim}$ is trivial. Since it was chosen arbitrarily, $\mathbf{Con}(\mathbf{A})/\overset{s}{\sim}$ is join semidistributive.

As noted when proving the dual result (Theorem 6.17) the implications (2) $\Longrightarrow$ (3) $\Longrightarrow$ (4) are straightforward.

We prove the contrapositive of (4) $\Longrightarrow$ (5). Suppose that there is an $\mathbf{L} \in \mathrm{CON}(\mathcal{V})$ that has a nonmodular $\mathrm{SD}_\vee$-failure. Then since $\mathrm{CON}(\mathcal{V})$ consists of the homomorphic images of lattices in the congruence prevariety of $\mathcal{V}$ there is an algebra $\mathbf{A} \in \mathcal{V}$, a sublattice $\mathbf{K} \leq \mathbf{Con}(\mathbf{A})$, and a surjective homomorphism $\sigma\colon \mathbf{K} \to \mathbf{L}$. The nonmodular $\mathrm{SD}_\vee$-failure I in $\mathbf{L}$ can be lifted to an $\mathrm{SD}_\vee$-failure $\overline{I}$ in $\mathbf{K}$ using the projectivity of the $\mathrm{SD}_\vee$-configuration. The fact that I is nonmodular means that it contains a sublattice isomorphic to the pentagon, $\mathbf{N}_5$. The pentagon is a projective lattice, so it can be lifted to a pentagon in $\overline{I}$, establishing that $\overline{I}$ is a nonmodular $\mathrm{SD}_\vee$-failure in $\mathbf{K}$, hence the interval in $\mathbf{Con}(\mathbf{A})$ generated by $\overline{I}$ is also a nonmodular $\mathrm{SD}_\vee$-failure. This proves that (4) $\Longrightarrow$ (5).

The implication (5) $\Longrightarrow$ (4) is trivial, so we complete the proof by showing that (4) $\Longrightarrow$ (1). Assume that (4) holds. This fact can be expressed by a quasi-identity valid in congruence lattices of $\mathcal{V}$. We have already discussed how to write down a quasi-identity for the dual property that $\mathrm{SD}_\wedge$-failures are modular in Remark 2.24, so it is easy to write down the one we need now. Namely, set $t := p \vee (q \wedge r)$, and for each i set $x_i^* := (x_i \wedge s) \vee t$. A quasi-identity expressing the fact that $\mathrm{SD}_\vee$-failures are modular is

$$\big((p \vee q) \approx s\big)\ \&\ \big((p \vee r) \approx s\big) \to u(x_1^*, x_2^*, x_3^*) \approx v(x_1^*, x_2^*, x_3^*) \tag{8.1}$$

where $u \approx v$ is the modular law. As explained in the paragraph after Definition 2.4, the fact that the premises of (8.1) are meet-free implies that (8.1) satisfies (W). By Theorem 2.22 this means that the class of varieties for which item (4) is true is definable by a set of idempotent Maltsev conditions. At least one of these must fail in the variety of semilattices, since, as we now prove, the variety of semilattices does not satisfy item (4). Let $\mathbf{2} = \langle\{0, 1\}; +\rangle$ be the 2-element join semilattice, and let $\mathbf{B} = \mathbf{2}^2$. We claim that $\mathbf{Con}(\mathbf{B}^2)$ has a nonmodular $\mathrm{SD}_\vee$-failure. To see this, let θ be the kernel of the homomorphism $+\colon \mathbf{B}^2 \to \mathbf{B}$. That is, $\theta = \{\langle (a, b), (c, d)\rangle \mid a + b = c + d\}$. Let $\eta_1 = \{\langle (a, b), (c, d)\rangle \mid a = c\}$ and $\eta_2 = \{\langle (a, b), (c, d)\rangle \mid b = d\}$ be the coordinate projection kernels. For any two pairs $(p, q), (r, s) \in B^2$ it is the case that

$$(p, q)\ \eta_1\ (p, p + q + r + s)\ \theta\ (p + q + r + s, p + q + r + s)\,,$$

and similarly $(p+q+r+s, p+q+r+s)\ \theta \circ \eta_1\ (r, s)$, so (p, q) is related to (r, s) by $\theta \vee \eta_1$. The pairs were arbitrary, so $\theta \vee \eta_1 = 1$; similarly $\theta \vee \eta_2 = 1$. This proves that $I[\theta \vee (\eta_1 \wedge \eta_2), 1] = I[\theta, 1]$ is an $\mathrm{SD}_\vee$-failure. But $\mathbf{B}^2/\theta$ is isomorphic to $\mathbf{B}$, since $+$ is a surjective homomorphism, so $I[\theta, 1] \cong \mathbf{Con}(\mathbf{B}) = \mathbf{Con}(\mathbf{2}^2) \cong \mathbf{D}_2$, which is a nonmodular lattice. This proves that any set of idempotent Maltsev conditions defining the class of varieties satisfying item (4) includes one that fails in the variety of semilattices, hence from Theorem 7.15 we get that any variety satisfying (4) must satisfy (1). □

The next result is an analogue of the $\mathbf{D}_1$-version of Theorem 4.23.

THEOREM 8.11. *Let $\mathcal{V}$ be a variety. The following conditions are equivalent.*

(1) *$\mathcal{V}$ satisfies an idempotent Maltsev condition that fails in the variety of semilattices.*
(2) *$\mathcal{V}$ satisfies a nontrivial congruence identity.*
(3) *$\mathbf{D}_2$ does not appear as a sublattice of $\mathbf{Con}(\mathbf{A})$ for any $\mathbf{A} \in \mathcal{V}$.*
(4) *$\mathbf{D}_2 \notin \mathrm{CON}(\mathcal{V})$.*

PROOF. We proved the equivalence of (1) and (2) in Theorem 7.15.

If item (3) holds, then $\mathcal{V}$ omits special $\mathbf{D}_2$'s. According to Theorem 5.28, this implies that $\mathcal{V}$ satisfies an idempotent Maltsev condition that fails in the variety of semilattices, so (3) $\Longrightarrow$ (1).

Now suppose that item (2) holds and that item (3) fails to hold. The failure of (3) implies that some $\mathbf{A} \in \mathcal{V}$ has a copy of $\mathbf{D}_2$ in its congruence lattice. Choose a copy and label it as in Figure 8.5. Then $\alpha \vee \beta = \alpha \vee \gamma$, and $\alpha \vee (\beta \wedge \gamma) = \alpha$.

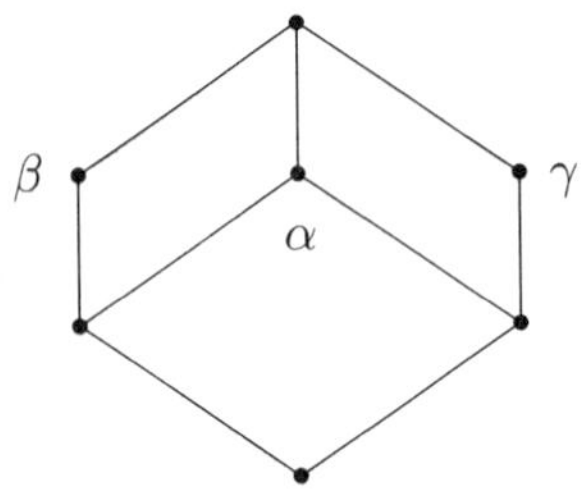

FIGURE 8.5

Therefore, since item (2) holds, Theorem 8.5 guarantees that $\alpha \lhd \alpha \vee \beta$. But this is forbidden by Theorem 4.16 (2). Thus (1), (2) and (3) are equivalent.

Both implications (4) $\Longrightarrow$ (2) and (4) $\Longrightarrow$ (3) are trivial, so to complete the proof we will show that items (2) and (3) jointly imply (4). Assume that items (2) and (3) are true for $\mathcal{V}$. From (2) and Theorem 8.3 we obtain that there is some positive integer M such that the identity $z^M \approx z^{M+1}$ holds in $\mathrm{CON}(\mathcal{V})$. Let $\mathcal{Z}^M$ denote the variety of lattices axiomatized by $z^M \approx z^{M+1}$. It is proved in Theorem 2.4 (5) $\Longrightarrow$ (6) of [48] that $\mathbf{D}_2$ is projective in any $\mathcal{Z}^M$ that contains $\mathbf{D}_2$. Since $\mathbf{D}_2$ is projective and subdirectly irreducible in $\mathcal{Z}^M$, the subclass $\mathcal{U} \subseteq \mathcal{Z}^M$ of lattices that have no sublattice isomorphic to $\mathbf{D}_2$ is a variety. By item (3), $\mathcal{U}$ contains the congruence lattices of members of $\mathcal{V}$, hence $\mathcal{U}$ contains the variety generated by these lattices. Since $\mathrm{CON}(\mathcal{V}) \subseteq \mathcal{U}$ and $\mathbf{D}_2 \notin \mathcal{U}$, (4) holds. □

The statement of the next theorem, which characterizes the class of varieties satisfying congruence identities in yet another way, should be compared to the statement of Theorem 4.12, which characterizes the class of varieties satisfying meet continuous congruence identities.

THEOREM 8.12. *The following conditions on a variety $\mathcal{V}$ are equivalent.*

(1) *$\mathcal{V}$ satisfies an idempotent Maltsev condition that fails in the variety of semilattices.*
(2) *The quasi-identity*

$$((p \vee q) \approx s) \,\&\, ((p \vee r) \approx s) \,\&\, (p^{[2]} \approx s) \to ((p \vee (q \wedge r)) \approx s) \tag{8.2}$$

holds in the congruence lattices of algebras in $\mathcal{V}$.
(3) *Quasi-identity (8.2) holds in* CON($\mathcal{V}$).
(4) *$\mathcal{V}$ satisfies a nontrivial congruence identity.*

Quasi-identity (8.2) in item (2) is the dual of quasi-identity (4.4) of Theorem 4.12 (2).

PROOF. We proved the equivalence of (1) and (4) in Theorem 7.15. The implication (3) $\Longrightarrow$ (2) is obvious. We have previously noted that $\mathbf{D}_1$ fails quasi-identity (4.4) of Theorem 4.12 (2), so of course $\mathbf{D}_2$ fails the dual quasi-identity of the present theorem. So, if the congruence lattices of algebras in $\mathcal{V}$ satisfy quasi-identity (8.2), then $\mathbf{D}_2$ is not embeddable in the congruence lattice of any algebra in $\mathcal{V}$. From Theorem 8.11 it follows that $\mathcal{V}$ satisfies a nontrivial congruence identity. This shows that (2) $\Longrightarrow$ (4). In summary, these observations show that (3) $\Longrightarrow$ (2) $\Longrightarrow$ (4) $\Longleftrightarrow$ (1). The rest of the proof will be devoted to showing that (4) $\Longrightarrow$ (3). We argue by contradiction, so assume that $\mathcal{V}$ satisfies a nontrivial congruence identity, but CON($\mathcal{V}$) fails to satisfy quasi-identity (8.2).

Quasi-identity (8.2) is equivalent modulo the axioms of lattice theory to the following quasi-identity:

$$((p \vee q) \approx s)\&((p \vee r) \approx s)\&(p^{[2]} \approx s)\&((q \wedge r) \leq p) \to (p \approx s)\,. \tag{8.3}$$

This is because any assignment of variables in some lattice that fails (8.3) also fails (8.2), while, if $p \mapsto a$, $q \mapsto b$, $r \mapsto c$, $s \mapsto d$ is an assignment that fails (8.2), then the reassignment $p \mapsto a' := a \vee (b \wedge c)$, $q \mapsto b$, $r \mapsto c$, $s \mapsto d$ clearly satisfies the first, second and fourth premise of (8.3) and fails the conclusion. To see that this reassignment also satisfies the third premise of (8.3) we must verify that $(a')^{[2]}(a', b, c) = d$. The fact that the first two premises are satisfied implies that d is the largest element of the sublattice generated by $\{a, b, c\}$. This shows that $(a')^{[2]}(a', b, c) \leq d$. But since the lattice operations are monotone, and $a \leq a'$, we get $d = a^{[2]}(a, b, c) \leq (a')^{[2]}(a', b, c)$. Thus $(a')^{[2]}(a', b, c) = d$, and the third premise is satisfied.

Let Q denote quasi-identity (8.3), and let $\big(\mathbf{P}(Q), p \approx s\big)$ be the Q-configuration. We have assumed that some $\mathbf{K} \in$ CON($\mathcal{V}$) fails Q, so there is a lattice $\mathbf{K} \in$ CON($\mathcal{V}$) and a homomorphism of partial lattices $\varphi\colon \mathbf{P}(Q) \to \mathbf{K}$ such that $\varphi(p) < \varphi(s)$. For concreteness, let $a := \varphi(p)$, $b := \varphi(q)$, $c := \varphi(r)$ and $d := \varphi(s)$. Since $\mathbf{K} \in$ CON($\mathcal{V}$) there is an algebra $\mathbf{A} \in \mathcal{V}$, a sublattice $\mathbf{L} \leq \mathbf{Con}(\mathbf{A})$ and a surjective homomorphism $\sigma\colon \mathbf{L} \to \mathbf{K}$. Choose a function $f\colon \{p, q, r, s\} \to \mathbf{L}$ such that $\sigma \circ f = \varphi$ on the domain of f. Since φ is a homomorphism of partial lattices and $q \wedge r \leq p, q, r, s$ in $\mathbf{P}(Q)$, the element $\varphi(q) \wedge \varphi(r) = b \wedge c$ is the smallest element of the sublattice generated by $\{a,b,c,d\}$. Thus, we may assume that f is chosen so that $f(q)\wedge f(r)\leq f(p), f(q), f(r), f(s)$.

The function f is an assignment of the variables of Q in $\mathbf{L}$. We wish to modify this assignment to a better one. Since $\mathbf{P}(Q)$ may be obtained (up to isomorphism) from $\mathbf{P}(\mathrm{SD}_\vee)$ by adding more relations, there is a (surjective) homomorphism $\psi\colon \mathbf{P}(\mathrm{SD}_\vee)\to\mathbf{P}(Q)$ that is the identity on the generators. Applying Lemma 8.4 to the assignment $f\circ\psi\colon \{p,q,r,s\}\to\mathbf{L}$ we obtain a homomorphism of partial lattices $\overline{\psi}\colon \mathbf{P}(\mathrm{SD}_\vee)\to\mathbf{L}$ such that $\sigma\circ\overline{\psi}=\varphi\circ\psi$ (Figure 8.6). The second

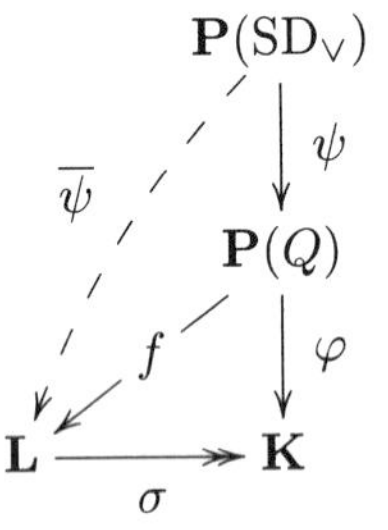

FIGURE 8.6

clause of Lemma 8.4 guarantees that

$$\begin{aligned}\overline{\psi}(q)\wedge\overline{\psi}(r) &\leq \big(f\circ\psi(q)\big)\wedge\big(f\circ\psi(r)\big)\\ &= f(q)\wedge f(r)\leq f(p)\\ &= f\circ\psi(p)\\ &\leq \overline{\psi}(p)\,,\end{aligned}\tag{8.4}$$

so if $\overline{\psi}$ is taken to define an assignment of the variables of Q in $\mathbf{L}$, $p\mapsto\alpha$, $q\mapsto\beta$, $r\mapsto\gamma$, $s\mapsto\delta$, then

(i) $\alpha\vee\beta=\delta$ and $\alpha\vee\gamma=\delta$ (by the projectivity of $\mathbf{P}(\mathrm{SD}_\vee)$),
(ii) $\beta\wedge\gamma\leq\alpha$ (by (8.4)), and
(iii) $\sigma(\alpha)=a<d=\sigma(\delta)$ (by the projectivity of $\mathbf{P}(\mathrm{SD}_\vee)$).

This assignment is not necessarily a failure of Q in $\mathbf{L}$, since the third premise $\alpha^{[2]}=\delta$ of Q may not be satisfied. But since the third premise was satisfied by $\varphi\colon \mathbf{P}(Q)\to\mathbf{K}$ we at least have

(iv) $\sigma(\alpha^{[2]})=a^{[2]}=d=\sigma(\delta)$.

The rest of the argument produces a contradiction from properties (i)–(iv). From items (i) and (ii) we derive that $I[\alpha,\alpha\vee\beta]$ is an $\mathrm{SD}_\vee$-failure, hence is an abelian interval according to Theorem 8.5. By this and item (i) we get $\alpha\vee\beta\overset{s}{\sim}\alpha\overset{s}{\sim}\alpha\vee\gamma$, which is hypothesis (1) of Lemma 6.27. Item (ii) above is the same as hypothesis (2) of Lemma 6.27, and our current assumption that $\mathcal{V}$ satisfies a congruence identity is strong enough to guarantee that $\mathcal{V}$ has a weak difference term. Since all the hypotheses of Lemma 6.27 currently hold, its conclusion also holds, which is the equality

$$\big(\beta\vee(\alpha\wedge\gamma)\big)\wedge\big(\gamma\vee(\alpha\wedge\beta)\big)=(\alpha\wedge\beta)\vee(\alpha\wedge\gamma)\,.$$

Using this to move from the first line to the second in

$$\begin{aligned}\alpha^{[2]}(\alpha,\beta,\gamma) &:= \alpha \vee \big((\beta \vee (\alpha \wedge \gamma)) \wedge (\gamma \vee (\alpha \wedge \beta))\big)\\ &= \alpha \vee \big((\alpha \wedge \beta) \vee (\alpha \wedge \gamma)\big)\\ &= \alpha\,,\end{aligned}$$

we derive that $\alpha^{[2]} = \alpha$. But from item (iii) we have $\sigma(\alpha) < \sigma(\delta)$, while from item (iv) we have $\sigma(\alpha^{[2]}) = \sigma(\delta)$, which together conflict with $\alpha^{[2]} = \alpha$. This contradiction completes the proof. □

The following result defines the class of varieties satisfying a nontrivial congruence identity by a 3-variable Maltsev condition.

THEOREM 8.13. *Let $\mathcal{V}$ be a variety. The following conditions are equivalent.*

(1) *$\mathcal{V}$ satisfies a nontrivial congruence identity.*
(2) $\mathcal{V} \models_{con} \alpha \cap (\beta \circ \gamma) \subseteq \big(\gamma \vee (\alpha \wedge \beta)\big) \circ \big(\beta \vee (\alpha \wedge \gamma)\big)$.
(3) *There is a positive integer k and ternary terms $d_0, \ldots, d_{2k+1}, e_0, \ldots, e_{2k+1}$, p such that $\mathcal{V}$ satisfies the following equations:*
   (i) $d_0(x,y,z) \approx p(x,y,z) \approx e_0(x,y,z)$;
   (ii) $d_i(x,y,y) \approx d_{i+1}(x,y,y)$ *and* $e_i(x,x,y) \approx e_{i+1}(x,x,y)$ *for even i;*
   (iii) $d_i(x,x,y) \approx d_{i+1}(x,x,y)$, $d_i(x,y,x) \approx d_{i+1}(x,y,x)$, $e_i(x,y,y) \approx e_{i+1}(x,y,y)$ *and* $e_i(x,y,x) \approx e_{i+1}(x,y,x)$ *for odd i;*
   (iv) $d_{2k+1}(x,y,z) \approx x$ *and* $e_{2k+1}(x,y,z) \approx z$.

PROOF. For locally finite varieties, this result is part of Theorem 9.8 of [**34**]. The proof given there shows that items (2) and (3) are equivalent for any variety, and that the idempotent Maltsev condition in item (3) fails in the variety of semilattices. Therefore (3) $\Longrightarrow$ (1), according to Theorem 7.15.

What remains to show is that (1) $\Longrightarrow$ (2). Let $\mathbf{F} = \mathbf{F}_{\mathcal{V}}(x,y,z)$ and set $\alpha := \mathrm{Cg}^{\mathbf{F}}(x,z)$, $\beta := \mathrm{Cg}^{\mathbf{F}}(x,y)$, and $\gamma := \mathrm{Cg}^{\mathbf{F}}(y,z)$. Since (x,z) is a generic element of $\alpha \cap (\beta \circ \gamma)$ it will suffice to prove that $(x,z) \in \big(\gamma \vee (\alpha \wedge \beta)\big) \circ \big(\beta \vee (\alpha \wedge \gamma)\big)$. Since $\alpha \vee \beta = \alpha \vee \gamma = \beta \vee \gamma =: \delta$, the intervals $I[\beta \vee (\alpha \wedge \gamma), \delta]$ and $I[\gamma \vee (\alpha \wedge \beta), \delta]$ are $\mathrm{SD}_{\vee}$-failures, and therefore $\beta \vee (\alpha \wedge \gamma) \overset{s}{\sim} \delta \overset{s}{\sim} \gamma \vee (\alpha \wedge \beta)$ (Theorem 8.5). Since these congruences are solvably related, they permute (Theorem 6.16). From $(x,y) \in \beta \vee (\alpha \wedge \gamma)$ and $(y,z) \in \gamma \vee (\alpha \wedge \beta)$ we get that

$$(x,z) \in \big(\beta \vee (\alpha \wedge \gamma)\big) \circ \big(\gamma \vee (\alpha \wedge \beta)\big) = \big(\gamma \vee (\alpha \wedge \beta)\big) \circ \big(\beta \vee (\alpha \wedge \gamma)\big).$$

□

8.3. Congruence Join Semidistributivity

The main purpose of this section is to prove that a variety is congruence join semidistributive if and only if it is both congruence meet semidistributive and satisfies a nontrivial congruence identity. Thus, congruence join semidistributive varieties are exactly those varieties whose members have no congruences that are 'abelian' in any of the senses that have been introduced.

THEOREM 8.14. *Let $\mathcal{V}$ be a variety. The following conditions are equivalent.*

(1) *$\mathcal{V}$ is congruence join semidistributive.*
(2) *If $\mathbf{A} \in \mathcal{V}$ and $\alpha, \beta, \gamma \in \mathrm{Con}(\mathbf{A})$, then*

$$\alpha \cap (\beta \circ \gamma) \subseteq \beta \vee (\alpha \wedge \gamma)\,.$$

(3) *There is a positive integer k and ternary terms $d_0, \ldots, d_k$ such that $\mathcal{V}$ satisfies the following equations:*
   (i) $d_0(x, y, z) \approx x$;
   (ii) $d_i(x, y, y) \approx d_{i+1}(x, y, y)$ *and* $d_i(x, y, x) \approx d_{i+1}(x, y, x)$ *for even* $i < k$;
   (iii) $d_i(x, x, y) \approx d_{i+1}(x, x, y)$ *for odd* $i < k$;
   (iv) $d_k(x, y, z) \approx z$.

(4) *$\mathcal{V}$ satisfies an idempotent Maltsev condition that fails in any nontrivial variety of modules and in the variety of semilattices.*

(5) *No member of $\mathcal{V}$ has a nontrivial abelian or rectangular congruence.*

(6) *$\mathcal{V}$ is congruence meet semidistributive and satisfies a nontrivial congruence identity.*

(7) *There is a positive integer N such that for every $\mathbf{A} \in \mathcal{V}$ and $\alpha, \beta, \gamma \in \mathrm{Con}(\mathbf{A})$,*

$$\alpha \vee (\beta \wedge \gamma) \approx (\alpha \vee \beta^N) \wedge (\alpha \vee \gamma^N).$$

(The lattice terms β^N and γ^N are defined in the paragraph preceeding Theorem 8.3.)

(8) *$\mathrm{CON}(\mathcal{V})$ is a join semidistributive variety of lattices.*

(9) $\mathbf{M}_3, \mathbf{D}_2 \notin \mathrm{CON}(\mathcal{V})$.

(10) $\mathbf{M}_3$ *and* $\mathbf{D}_2$ *are not embeddable in* $\mathbf{Con}(\mathbf{A})$ *for any* $\mathbf{A} \in \mathcal{V}$.

PROOF. We prove (1)–(6) cyclically, and then show the implications ((1) & (6)) $\implies$ (7) $\implies$ (8) $\implies$ (9) $\implies$ (10) $\implies$ (6).

Assume that (1) holds and that $\mathbf{A} \in \mathcal{V}$ has congruences α, β and γ. Choose any $(a, c) \in \alpha \cap (\beta \circ \gamma)$. Then $a \equiv_\beta b \equiv_\gamma c$ for some $b \in A$. Let $\alpha' = \mathrm{Cg}^{\mathbf{A}}(a, c)$, $\beta' = \mathrm{Cg}^{\mathbf{A}}(a, b)$ and $\gamma' = \mathrm{Cg}^{\mathbf{A}}(b, c)$. Since the generating pairs are related by α, β and γ respectively, we get that $\alpha' \leq \alpha$, $\beta' \leq \beta$ and $\gamma' \leq \gamma$. The congruence $\beta' \vee \alpha'$ is generated by (a, c) and (a, b), so it is the congruence on $\mathbf{A}$ generated by $X \times X$ where $X = \{a, b, c\}$. Similarly $\beta' \vee \gamma'$ is the congruence generated by $X \times X$, so $\beta' \vee \alpha' = \beta' \vee \gamma'$. By (1), we get that this congruence is also $\beta' \vee (\alpha' \wedge \gamma')$. Since this congruence contains (a, c), we get that

$$(a, c) \in \beta' \vee (\alpha' \wedge \gamma') \leq \beta \vee (\alpha \wedge \gamma),$$

where the inequality follows from the monotonicity of the lattice operations. This argument shows that $(a, c) \in \beta \vee (\alpha \wedge \gamma)$ for every $(a, c) \in \alpha \cap (\beta \circ \gamma)$, so $\alpha \cap (\beta \circ \gamma) \subseteq \beta \vee (\alpha \wedge \gamma)$ and (2) holds.

For (2) $\implies$ (3), let $p = \alpha \cap (\beta \circ \gamma)$, $r = \beta \vee (\alpha \wedge \gamma)$, and let r_i be the $\{\circ, \wedge\}$-word obtained from r by replacing $\vee$ with i-fold relational product. Then the congruence inclusion of (2) has the form $p \subseteq \bigcup_{i \in \omega} r_i$ with the r_i's increasing. By the remark after Theorem 4.7, such a congruence inclusion is associated to a Maltsev condition. The method described in the proof of Theorem 4.7 for producing the Maltsev condition yields the one in item (3).

To prove that (3) $\implies$ (4), we will show that the idempotent Maltsev condition in (3) fails in every nontrivial variety of modules and in the variety of semilattices. Assume that $\mathcal{M}$ is the variety of all modules over the ring $\mathbf{R}$, and that $\mathcal{M}$ satisfies the Maltsev condition in item (3). Each d_i is a module term, so $d_i(x, y, z) = a_i x + b_i y + c_i z$ for some $a_i, b_i, c_i \in R$. Since $d_i(x, x, x) \approx x$, we get that $a_i x + b_i x + c_i x \approx (a_i + b_i + c_i)x \approx x$ holds in any $\mathbf{M} \in \mathcal{M}$, hence $a_i + b_i + c_i = 1$ in $\mathbf{R}$. The first equation in (3)(ii) implies that $a_i = a_{i+1}$ for even i, the second equation in (3)(ii) implies that $b_i = b_{i+1}$ for even i, and using the equations of the type $a_i + b_i + c_i = 1$ we see that $c_i = c_{i+1}$ holds for all even i. The equation from (3)(iii) implies that $c_i = c_{i+1}$ holds for odd i, so $c_i = c_{i+1}$ for all i. But the equation from (3)(i) implies that $c_0 = 0$, while the equation from (3)(iv) implies that $c_k = 1$. Altogether we derive that $0 = 1$ in $\mathbf{R}$, so $\mathcal{M}$ is a trivial variety of modules.

Now suppose that $\mathcal{S}$ is the variety of semilattices, and that $\mathcal{S}$ satisfies the Maltsev condition in (3). Choose k minimal for the condition in (3) to be satisfied in $\mathcal{S}$. By equations (3)(i) and (3)(iv), it must be that $k > 0$. If $k > 0$ is even, then by equation (3)(iii) we find that $d_{k-1}(x,x,y) \approx d_k(x,x,y) \approx y$, which for semilattices implies that $d_{k-1}(x,y,z) \approx z$. Thus we can delete the term d_k and have a shorter sequence of terms satisfying the conditions in (3). If $k > 0$ is odd, then the equations in (3)(ii) imply that d_{k-1} does not depend on its first variable (since $d_{k-1}(x,y,y) \approx d_k(x,y,y) \approx y$), and does not depend on its second variable (since $d_{k-1}(x,y,x) \approx d_k(x,y,x) \approx x$). Therefore $d_{k-1}(x,y,z) \approx d_{k-1}(z,z,z) \approx z$, and we can again shorten the sequence. Hence the equations in (3) cannot be satisfied by semilattice terms. Thus (3) $\Longrightarrow$ (4).

It follows from Theorem 8.1 (10) $\Longrightarrow$ (4) that if $\mathcal{V}$ satisfies an idempotent Maltsev condition that fails in every nontrivial variety of modules, then $\mathcal{V}$ omits abelian congruences. It follows from Theorem 5.25 that if $\mathcal{V}$ satisfies an idempotent Maltsev condition that fails in the variety of semilattices, then $\mathcal{V}$ omits rectangular tolerances (hence $\mathcal{V}$ omits rectangular congruences by Theorem 5.22). Thus (4) $\Longrightarrow$ (5).

If (5) holds, then $\mathcal{V}$ has no abelian congruence, so from (4) $\Longrightarrow$ (1) of Theorem 8.1 we derive that $\mathcal{V}$ is congruence meet semidistributive. The variety $\mathcal{V}$ also has no rectangular congruence, or tolerance, so by Theorems 5.25 and 7.15 we have that $\mathcal{V}$ satisfies a nontrivial congruence identity. Hence (6) holds.

Now suppose that $\mathcal{V}$ is congruence meet semidistributive and satisfies a nontrivial congruence identity. The latter supposition implies that the $\mathrm{SD}_\vee$-failures in congruence lattices of members of $\mathcal{V}$ are all abelian, according to Theorem 8.5. But the former supposition implies that there are no abelian intervals in congruence lattices of members of $\mathcal{V}$, according to (1) $\Longrightarrow$ (4) of Theorem 8.1. This shows that there are no $\mathrm{SD}_\vee$-failures at all, which means of course that $\mathcal{V}$ is congruence join semidistributive.

Now that the first six items of the theorem have been shown to be equivalent, we turn to the seventh. Assume that (1) and (the equivalent property) (6) hold. From (6) and Theorem 8.3 we know that there is a positive integer M such that the identity

$$z^M(x,y,z) \approx z^{M+1}(x,y,z)$$

holds in all congruence lattices of members of $\mathcal{V}$. Thus, if α, β and γ satisfy the conditions in the statement of item (7), and $\beta^k := y^k(\alpha,\beta,\gamma)$ and $\gamma^k := z^k(\alpha,\beta,\gamma)$, then $\beta^M = \beta^{M+1}$ and $\gamma^M = \gamma^{M+1}$. This implies that $\alpha \vee \beta^M \geq \alpha \vee \gamma^{M+1} = \alpha \vee \gamma^M$, and similarly that $\alpha \vee \gamma^M \geq \alpha \vee \beta^M$, so $\alpha \vee \beta^M = \alpha \vee \gamma^M$. Since (1) holds, we have

$$\alpha \vee (\beta^M \wedge \gamma^M) = \alpha \vee \beta^M .$$

But $\beta \wedge \gamma \leq \beta^M \wedge \gamma^M \leq \beta^M \leq \beta$ and $\beta \wedge \gamma \leq \beta^M \wedge \gamma^M \leq \gamma^M \leq \gamma$, so $\beta^M \wedge \gamma^M = \beta \wedge \gamma$. Therefore

$$\alpha \vee (\beta \wedge \gamma) = \alpha \vee \beta^M ,$$

and similarly

$$\alpha \vee (\beta \wedge \gamma) = \alpha \vee \gamma^M .$$

This proves that $\alpha \vee (\beta \wedge \gamma) = (\alpha \vee \beta^M) \wedge (\alpha \vee \gamma^M)$ for the fixed value of M provided by Theorem 8.3.

If item (7) holds, then $\mathrm{CON}(\mathcal{V})$ satisfies the weakened distributive law $x \vee (y \wedge z) \approx (x \vee y^N) \wedge (x \vee z^N)$ for some N. We claim that any lattice satisfying this law is join semidistributive. Indeed, if $\mathbf{L}$ is a lattice, $a, b, c \in L$, and $b^k := y^k(a, b, c)$ and $c^k := z^k(a, b, c)$, then it may be shown by induction that if $a \vee b = a \vee c$, then $b^k = b$ and $c^k = c$ for all k. Thus, if $d := a \vee b = a \vee c$, the identity in (7) implies that

$$a \vee (b \wedge c) = (a \vee b^N) \wedge (a \vee c^N) = (a \vee b) \wedge (a \vee c) = d .$$

Since $a, b, c \in L$ were chosen arbitrarily, $\mathbf{L}$ is join semidistributive.

Item (8) implies item (9) because $\mathbf{M}_3$ and $\mathbf{D}_2$ are not join semidistributive.

Item (9) implies item (10) because sublattices of $\mathbf{Con}(\mathbf{A})$ belong to $\mathrm{CON}(\mathcal{V})$.

Assume that item (10) holds. Since $\mathbf{M}_3$ is not embeddable in the congruence lattice of any member of $\mathcal{V}$, it follows from Theorem 8.1 that $\mathcal{V}$ is congruence meet semidistributive. Since $\mathbf{D}_2$ is not embeddable in the congruence lattice of any member of $\mathcal{V}$, it follows from Theorem 8.11 that $\mathcal{V}$ satisfies a nontrivial congruence identity. Thus item (6) holds. The proof is complete. □

This theorem provides a positive solution to Problem 2.18 of [**40**] from this theorem, which we extract as the following corollary.

COROLLARY 8.15. *The class of congruence join semidistributive varieties is definable by a(n idempotent) Maltsev condition.*

PROOF. By Theorem 8.14 (1) $\Longleftrightarrow$ (3). □

CHAPTER 9

Residually Small Varieties

A variety is **residually large** if it has a proper class of subdirectly irreducible members up to isomorphism, otherwise it is **residually small**. Our primary goal in this chapter is to prove that a residually small variety satisfies a congruence identity if and only if it is congruence modular. (David Hobby and Ralph McKenzie proved essentially the same result for locally finite varieties in Chapter 10 of [**34**].) Our result nearly completes the classification of congruence varieties associated to residually small varieties. In Section 9.2, we apply the result to show that almost congruence distributive varieties cannot have a Taylor term.

9.1. Residual Smallness and Congruence Modularity

LEMMA 9.1. *Let $\mathcal{V}$ be a variety that satisfies a nontrivial congruence identity. If $\mathbf{A} \in \mathcal{V}$ has congruences α, β and γ that generate a sublattice isomorphic to $\mathbf{N}_5$ in* Con($\mathbf{A}$)*, as depicted in Figure* 9.1*, then*

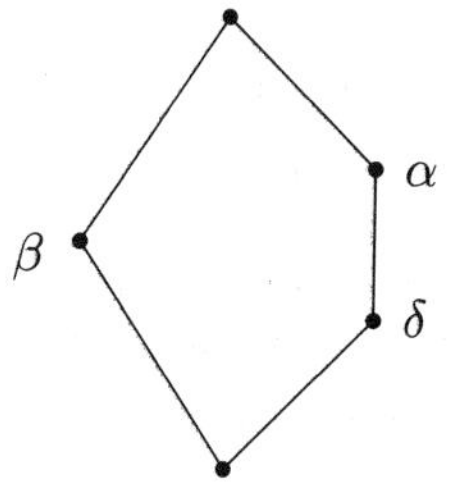

FIGURE 9.1

(1) $\mathbf{C}(\alpha, \beta; \delta)$ *holds, and*
(2) $\mathbf{C}(\beta, \alpha; \delta)$ *fails in the following strong way: There exists a $(\beta, \beta; 0)$-pair (a, b) (cf. Definition* 3.14*), a polynomial $f(x, \mathbf{y})$ and tuples $\mathbf{c}\ \alpha\ \mathbf{d}$ such that*

$$f(b, \mathbf{c}) \equiv_\delta f(b, \mathbf{d}),$$

but

$$f(a, \mathbf{c}) \not\equiv_\delta f(a, \mathbf{d}).$$

PROOF. Item (1) follows from Theorem 2.19 (8), since

$$\beta \wedge \big(\alpha \vee (\beta \wedge \delta)\big) \leq \delta .$$

For (2), assume instead that

$$f(b, \mathbf{c}) \equiv_\delta f(b, \mathbf{d}) \implies f(a, \mathbf{c}) \equiv_\delta f(a, \mathbf{d})$$

holds for every $(\beta,\beta;0)$-pair (a,b). Then by transitivity this implication holds whenever $a \geq_\beta b$. But since $\mathcal{V}$ has no algebra with a rectangular tolerance, Theorem 5.23 (1) $\iff$ (4) guarantees that the set of pairs (a,b) satisfying $a \geq_\beta b$ contains β. Thus, if this implication holds for every f and every choice of tuples $\mathbf{c}\ \alpha\ \mathbf{d}$ whenever $(a,b) \in \beta$, then $\mathbf{C}(\beta,\alpha;\delta)$ holds. But $\mathbf{C}(\beta,\alpha;\delta)$ does not hold, as we proved in Theorem 4.16 (2). Therefore f, $\mathbf{c}$ and $\mathbf{d}$ exist. $\square$

As it happens, the properties described in conditions (1) and (2) of Lemma 9.1 are all that are needed to establish the existence of a proper class of subdirectly irreducibles in $\mathcal{V}$. (We no longer need the assumption that $\mathcal{V}$ satisfies a nontrivial congruence identity.)

THEOREM 9.2. *Let $\mathcal{V}$ be a variety. Assume that $\mathbf{A} \in \mathcal{V}$ has congruences $\delta < \alpha$ and a tolerance T such that*

(1) *$\mathbf{C}(\alpha, T;\delta)$ holds, and*
(2) *For some $(T,T;0)$-pair (a,b) there is a polynomial $f(x,\mathbf{y})$ and tuples $\mathbf{c}\ \alpha\ \mathbf{d}$ such that*
$$v := f(b,\mathbf{c}) \equiv_\delta f(b,\mathbf{d}) =: v'\,,$$
but
$$u := f(a,\mathbf{c}) \not\equiv_\delta f(a,\mathbf{d}) =: u'\,.$$

Then $\mathcal{V}$ is residually large.

PROOF. Both (1) and (2) continue to hold if we factor by δ, so there is no loss in generality in assuming that $\delta = 0$. (Under this assumption $v = v'$.) We now explain how to construct a subdirectly irreducible algebra in $\mathcal{V}$ of cardinality $\geq \kappa$ for any infinite cardinal κ.

Let $\mathbf{B}$ be the subalgebra of $\mathbf{A}^\kappa$ consisting of those κ-tuples $\mathbf{a}$ where $(a_i,a_j) \in T$ for all coordinates $i < j < \kappa$. For any $a \in A$ let $\widehat{a}$ denote the constant κ-tuple satisfying $(\widehat{a})_i = a$ for all i. Clearly $\widehat{a} \in B$ for all $a \in A$. Let $N = \{u,v\}$ and let $N' = \{u',v'\}$. Since $(a,b) \in T$ it follows that $(u,v) = \big(f(a,\mathbf{c}), f(b,\mathbf{c})\big) \in T$ and $(u',v') \in T$, so $N^\kappa \cup (N')^\kappa \subseteq B$. For an element $\mathbf{z}$ of N^κ denote by $\mathbf{z}'$ the element of $(N')^\kappa$ obtained from $\mathbf{z}$ by priming the entry in each coordinate. Furthermore, let
$$G = \big\{(\mathbf{z},\mathbf{z}') \in B^2 \mid \mathbf{z} \in N^\kappa \setminus \{\widehat{u}\}\big\}\,,$$
and let $\psi = \mathrm{Cg}^{\mathbf{B}}(G)$.

CLAIM 9.3. *The pair $(\hat{u},\hat{u}')$ is not contained in ψ.*

As $u \neq u'$, it is sufficient to prove that $\{\widehat{u}\}$ is a singleton ψ-class. For this we must show that if $(\mathbf{z},\mathbf{z}') \in G$ and p is a unary polynomial of $\mathbf{B}$, then
$$p(\mathbf{z}) = \widehat{u} \iff p(\mathbf{z}') = \widehat{u}\,.$$
Since $\mathbf{z} \in N^\kappa \setminus \{\widehat{u}\}$, at least one coordinate of $\mathbf{z}$ is v. If $z_i = v$, then $z_i' = v' = v$, so $\mathbf{z}$ and $\mathbf{z}'$ agree in the i-th coordinate. This implies that $p(\mathbf{z})$ and $p(\mathbf{z}')$ agree in the i-th coordinate. We now argue that these tuples also agree in the j-th coordinate for every other $j < \kappa$. There is nothing to prove for those coordinates where $z_j = v = v' = z_j'$, by the above argument, so we consider only those j where $z_j = u$ and $z_j' = u'$.

The polynomial p of $\mathbf{B}$ comes from a term $g(x,\mathbf{y})$, where appropriate parameters from $\mathbf{B}$ are substituted in place of $\mathbf{y}$ in every coordinate. Denote by $\mathbf{r}$ and $\mathbf{s}$

the parameters occurring at the i-th and the j-th coordinate, respectively. Then $\mathbf{r}\ T\ \mathbf{s}$ holds by the definition of $\mathbf{B}$. We have to prove that $g(u, \mathbf{s}) = g(u', \mathbf{s})$, and we have either

$$g(u, \mathbf{s}) = u = g(v, \mathbf{r}) \tag{9.1}$$

(in case $p(\mathbf{z}) = \widehat{u}$), or

$$g(u', \mathbf{s}) = u = g(v', \mathbf{r}) \tag{9.2}$$

(in case $p(\mathbf{z}') = \widehat{v}$). However $C(\alpha, T; 0)$ holds and $\mathbf{c}\ \alpha\ \mathbf{d}$, so

$$\big(g(u, \mathbf{s}) = \big) \quad g\big(f(a, \underline{\mathbf{c}}), \mathbf{s}\big) = g\big(f(b, \underline{\mathbf{c}}), \mathbf{r}\big) \quad \big(= g(v, \mathbf{r})\big)$$

is equivalent to

$$\big(g(u', \mathbf{s}) = \big) \quad g\big(f(a, \underline{\mathbf{d}}), \mathbf{s}\big) = g\big(f(b, \underline{\mathbf{d}}), \mathbf{r}\big) \quad \big(= g(v', \mathbf{r})\big).$$

Since at least one of (9.1) or (9.2) holds, we get that $g(u, \mathbf{s}) = g(v, \mathbf{r})$ and $g(u', \mathbf{s}) = g(v', \mathbf{r})$ both hold. Moreover, $v = v'$ implies that $g(v, \mathbf{r}) = g(v', \mathbf{r})$, so in fact

$$g(u, \mathbf{s}) = g(v, \mathbf{r}) = g(v', \mathbf{r}) = g(u', \mathbf{s}).$$

Thus $g(u, \mathbf{s}) = g(u', \mathbf{s})$ indeed holds, completing the argument that $p(\mathbf{z}) = \widehat{u} \iff p(\mathbf{z}') = \widehat{u}$ holds. This proves the claim.

Since $(\widehat{u}, \widehat{u}') \notin \psi$ we may extend ψ to a congruence $\psi_0 \geq \psi$ that is maximal with respect to $(\widehat{u}, \widehat{u}') \notin \psi_0$. Then $\mathbf{B}/\psi_0$ is a subdirectly irreducible algebra in $\mathcal{V}$. We claim that $|B/\psi_0| \geq \kappa$.

In order to prove that $\mathbf{B}/\psi_0$ is this large, we need only to exhibit κ elements of $\mathbf{B}$ that are pairwise incongruent modulo ψ_0. We introduce the following notation for certain elements of A^κ: for $i < j < \kappa$, and for arbitrary elements $a, b, c \in A$ let $[a, b, c]^{i,j}$ denote the tuple whose k-th coordinate is a for $0 \leq k < i$, is b for $i \leq k < j$, and is c for $j \leq k < \kappa$. (Note that $[a, b, c]^{i,j} \in B$ if a, b and c are pairwise T-related.) Similarly, if $\mathbf{a}, \mathbf{b}, \mathbf{c} \in A^n$, then let $[\mathbf{a}, \mathbf{b}, \mathbf{c}]^{i,j}$ denote the member of B^n whose ℓ-th coordinate is $[a_\ell, b_\ell, c_\ell]^{i,j}$ for $1 \leq \ell \leq n$. We shall omit the superscripts i, j if we think it is clear from the context what they are.

As (a, b) is a $(T, T; 0)$-pair, there exists a T, T-matrix

$$\begin{bmatrix} f(\mathbf{p}, \mathbf{r}) & f(\mathbf{p}, \mathbf{s}) \\ f(\mathbf{q}, \mathbf{r}) & f(\mathbf{q}, \mathbf{s}) \end{bmatrix} = \begin{bmatrix} * & a \\ a & b \end{bmatrix}.$$

CLAIM 9.4. *For each $0 < i < j < \kappa$ the tuples $[\mathbf{q}, \mathbf{p}, \mathbf{p}]$ and $[\mathbf{q}, \mathbf{q}, \mathbf{p}]$ are not ψ_0-congruent in every coordinate.*

Suppose that they are. Then in the algebra $\mathbf{B}$ we have that

$$f([\mathbf{q}, \mathbf{p}, \mathbf{p}], [\mathbf{r}, \mathbf{s}, \mathbf{s}])\ \psi_0\ f([\mathbf{q}, \mathbf{q}, \mathbf{p}], [\mathbf{r}, \mathbf{s}, \mathbf{s}]),$$

that is,

$$[a, a, a]\ \psi_0\ [a, b, a].$$

Therefore

$$[u, u, u] = f([a, a, a], [\mathbf{c}, \mathbf{c}, \mathbf{c}])\ \psi_0\ f([a, b, a], [\mathbf{c}, \mathbf{c}, \mathbf{c}]) = [u, v, u],$$

and

$$[u', u', u'] = f([a, a, a], [\mathbf{d}, \mathbf{d}, \mathbf{d}])\ \psi_0\ f([a, b, a], [\mathbf{d}, \mathbf{d}, \mathbf{d}]) = [u', v', u'].$$

But then

$$\widehat{u} = [u, u, u]\ \psi_0\ [u, v, u]\ \psi\ [u', v', u']\ \psi_0\ [u', u', u'] = \widehat{u}'.$$

By transitivity and the fact that $\psi_0 \geq \psi$, this yields $(\widehat{u}, \widehat{u}') \in \psi_0$, contradicting the choice of ψ_0. So the claim is proved.

To finish the proof, let m be the length of the tuples $\mathbf{p}$ and $\mathbf{q}$, and define a mapping $h\colon \kappa \to (B/\psi_0)^m$ in the following way. Let $i < k < \kappa$, and

$$h(i) = [\mathbf{q}, \mathbf{p}, \mathbf{p}]^{i,k}/\psi_0^m\,.$$

The value of $h(i)$ does not depend on k because the last two blocks of coordinates of the tuple $[\mathbf{q}, \mathbf{p}, \mathbf{p}]$ are both $\mathbf{p}$. We claim that h is injective. Indeed, suppose that $i < j < \kappa$. Then

$$h(i) = [\mathbf{q}, \mathbf{p}, \mathbf{p}]^{i,j}/\psi_0^m$$

and

$$h(j) = [\mathbf{q}, \mathbf{q}, \mathbf{p}]^{i,j}/\psi_0^m\,,$$

and these tuples are different by Claim 9.4. Therefore h is indeed injective, so $\kappa \leq |B/\psi_0|^m = |B/\psi_0|$, proving the theorem. □

We obtain the main theorem of this chapter as a combination of Lemma 9.1 and Theorem 9.2.

THEOREM 9.5. *A residually small variety satisfies a congruence identity if and only if it is congruence modular.* □

This theorem generalizes Theorem 10.4 of [**34**] from locally finite varieties to arbitrary varieties.

REMARK 9.6. We describe a three-step plan for classifying all congruence varieties associated to residually small varieties.

(1) Prove that if $\mathcal{V}$ is residually small and satisfies a nontrivial congruence identity, then $\mathcal{V}$ is congruence modular.
(2) Prove that if $\mathcal{V}$ is residually small and congruence modular, then $\mathcal{V}$ is congruence distributive or $\mathrm{CON}(\mathcal{V}) = \mathrm{CON}(\mathcal{M})$ for some variety $\mathcal{M}$ of modules.
(3) Classify the congruence varieties of varieties of modules.

Theorem 9.5 accomplishes Step (1). Step (3) was accomplished by Gábor Czédli and George Hutchinson in [**36**]. Partial results on Step (2) were obtained by Alan Day and Emil W. Kiss in [**12**]. Namely, Day and Kiss accomplished Step (2) under the further assumption that $\mathcal{V}$ is either locally finite or locally solvable. Their ideas extend a little further than these two cases, but fall short of fully accomplishing Step (2). Therefore, we pose the problem of accomplishing Step (2) in full generality.

PROBLEM 9.7. Show that if $\mathcal{V}$ is residually small and congruence modular, then $\mathcal{V}$ is congruence distributive or $\mathrm{CON}(\mathcal{V}) = \mathrm{CON}(\mathcal{M})$ for some variety $\mathcal{M}$ of modules.

If this is proved, then we will know that the congruence varieties $\mathrm{CON}(\mathcal{V})$ of residually small varieties $\mathcal{V}$ are precisely

- the variety of all lattices — e.g. when $\mathcal{V}$ is the (residually small) variety of sets,
- $\mathrm{CON}(\mathcal{M})$ where $\mathcal{M}$ is a variety of modules,
- the variety of distributive lattices — e.g. when $\mathcal{V}$ is the (residually small) variety of distributive lattices, and
- the variety of trivial lattices (e.g. when $\mathcal{V}$ is a trivial variety).

9.2. Almost Congruence Distributive Varieties

A variety is **affine complete** if every congruence preserving function on every member is a polynomial. In [**43**], Kalle Kaarli and Alden Pixley announced that any affine complete variety is residually finite and has the property that every finitely generated subvariety is congruence distributive. At the time it was not known whether the entire variety had to be congruence distributive. Pixley then introduced the following concept.

DEFINITION 9.8. A variety $\mathcal{V}$ is called **almost congruence distributive**, or **ACD**, if it is residually finite, every finitely generated subvariety is congruence distributive, but $\mathcal{V}$ itself is not congruence distributive.

Pixley asked if there exists an ACD variety. A negative answer would, of course, show that every affine complete variety is congruence distributive.

Kaarli and Ralph McKenzie proved in [**42**] that every affine complete variety is congruence distributive, but it remains open whether there is an ACD variety. This is still an interesting question, since we know very little about how local properties of a variety, like the structure of the subdirectly irreducible algebras, imply global properties, such as the satisfaction of a Maltsev condition.

In this section we prove that an ACD variety satisfies no nontrivial idempotent Maltsev condition. This implies that if an ACD variety exists, then some member has a nontrivial strongly abelian congruence, but for some reason such congruences do not show up in any subdirectly irreducible algebra or even in any subvariety generated by a finite set of subdirectly irreducible algebras.

THEOREM 9.9. *Let $\mathcal{V}$ be a variety satisfying a nontrivial idempotent Maltsev condition.*

(1) *If every subdirectly irreducible algebra in $\mathcal{V}$ generates a subvariety satisfying a nontrivial congruence identity, then $\mathcal{V}$ satisfies a nontrivial congruence identity.*
(2) *If every subdirectly irreducible algebra in $\mathcal{V}$ generates a congruence join semidistributive subvariety, then $\mathcal{V}$ is congruence join semidistributive.*

Now assume also that $\mathcal{V}$ is residually small.

(1)′ *If every subdirectly irreducible algebra in $\mathcal{V}$ generates a subvariety satisfying a nontrivial congruence identity, then $\mathcal{V}$ is congruence modular.*
(2)′ *If every subdirectly irreducible algebra in $\mathcal{V}$ generates a congruence join semidistributive subvariety, then $\mathcal{V}$ is congruence distributive.*

PROOF. If each subdirectly irreducible algebra in $\mathcal{V}$ generates a subvariety satisfying a nontrivial congruence identity, then by Theorem 7.15 each subdirectly irreducible algebra in $\mathcal{V}$ has a Hobby–McKenzie term. By Theorem 5.25 the rectangular tolerances on subdirectly irreducible members of $\mathcal{V}$ are trivial. Corollary 5.20 guarantees that no algebra in $\mathcal{V}$ has a rectangular tolerance. Reusing Theorems 5.25 and 7.15, we get that $\mathcal{V}$ satisfies a nontrivial congruence identity. This proves (1).

For (2), if every subdirectly irreducible in $\mathcal{V}$ generates a congruence join semidistributive subvariety, then no subdirectly irreducible algebra has a nontrivial rectangular or abelian tolerance. By part (1), this implies that $\mathcal{V}$ satisfies a nontrivial congruence identity, hence $\mathcal{V}$ has a weak difference term according to Corollary 6.3. This shows that the hypotheses of Corollary 6.9 are met, so from the fact that no subdirectly irreducible algebra has an abelian tolerance we may derive that no algebra in $\mathcal{V}$ has an abelian tolerance. By Theorem 8.14 (5) $\Longrightarrow$ (1) it follows that $\mathcal{V}$ is congruence join semidistributive.

Items (1)′ and (2)′ follow from items (1) and (2) using Theorem 9.5. □

COROLLARY 9.10. *An ACD variety satisfies no nontrivial idempotent Maltsev condition.*

PROOF. This follows from Theorem 9.9 (2)′ and the definition of an ACD variety. □

Theorem 9.9 suggests the following problem.

PROBLEM 9.11. Suppose that $\mathcal{V}$ satisfies a nontrivial idempotent Maltsev condition, and every subdirectly irreducible member of $\mathcal{V}$ generates a congruence modular subvariety. Must $\mathcal{V}$ be congruence modular? (We are interested in the case where $\mathcal{V}$ is residually large.)

The answer to Problem 9.11 is affirmative if $\mathcal{V}$ is locally finite. Indeed, it follows from tame congruence theory that the following conditions are equivalent for a locally finite variety:

(1) $\mathcal{V}$ is congruence modular.
(2) The minimal sets of the finite algebras are of type $\mathbf{2}, \mathbf{3}$ or $\mathbf{4}$ and have empty tails.
(3) The minimal sets of the finite subdirectly irreducible algebras are of type $\mathbf{2}, \mathbf{3}$ or $\mathbf{4}$ and have empty tails.
(4) Each finite subdirectly irreducible algebra generates a congruence modular subvariety.

However, the modular law is special in this regard, because of its link with the empty tails condition. We expect that for identities weaker than modularity the problem analogous to Problem 9.11 will have a negative answer.

Another problem is suggested by the proof of Theorem 9.9.

PROBLEM 9.12. Suppose that $\mathcal{V}$ has a Taylor term. If some algebra in $\mathcal{V}$ has a nontrivial abelian tolerance, must some subdirectly irreducible algebra in $\mathcal{V}$ have a nontrivial abelian tolerance?

Problems

In this aftersection, we collect all the problems from the text and list a few more. To learn the context of the first set of problems, the reader should revisit that part of the text. In the second set we include a short explanation of the background of each problem.

From the text

PROBLEM 4.1. Which congruence prevarieties are first-order axiomatizable?

PROBLEM 4.2. Is the congruence prevariety of a congruence modular variety first-order axiomatizable?

PROBLEM 4.3. Is the class of varieties satisfying a given set of congruence quasi-identities definable by (idempotent, linear) Maltsev conditions?

PROBLEM 4.4. Suppose that a variety $\mathcal{V}$ satisfies no nontrivial idempotent Maltsev condition. Is it true that $\mathcal{L}(\mathcal{V}) = \mathcal{L}$? Is it true at least that $\mathcal{L}(\mathcal{V})$ contains the lattice Π_4 of all equivalence relations on a 4-element set?

PROBLEM 4.6. Determine the number of meet continuous lattice varieties.

PROBLEM 4.14. Is there a weakest congruence quasi-identity?

PROBLEM 6.21. Assume that $\mathcal{V}$ has a weak difference term, and that $\mathbf{A} \in \mathcal{V}$ is ∞-solvable. Does $\mathbf{A}$ have a local Maltsev term operation? Does $\mathbf{A}$ have a Maltsev term operation? If $\mathcal{V}$ is generated by ∞-solvable algebras, then must $\mathcal{V}$ be congruence permutable?

PROBLEM 9.7. Show that if $\mathcal{V}$ is residually small and congruence modular, then $\mathcal{V}$ is congruence distributive or $\mathrm{CON}(\mathcal{V}) = \mathrm{CON}(\mathcal{M})$ for some variety $\mathcal{M}$ of modules.

PROBLEM 9.11. Suppose that $\mathcal{V}$ satisfies a nontrivial idempotent Maltsev condition, and every subdirectly irreducible member of $\mathcal{V}$ generates a congruence modular subvariety. Must $\mathcal{V}$ be congruence modular? (We are interested in the case where $\mathcal{V}$ is residually large.)

PROBLEM 9.12. Suppose that $\mathcal{V}$ has a Taylor term. If some algebra in $\mathcal{V}$ has a nontrivial abelian tolerance, must some subdirectly irreducible algebra in $\mathcal{V}$ have a nontrivial abelian tolerance?

Additional problems

PROBLEM P1. Suppose that $\mathcal{V}$ satisfies no congruence identity. Is it true that the congruence prevariety of $\mathcal{V}$ contains the congruence prevariety of the variety of semilattices? What if $\mathcal{V}$ has a join term?

If $\mathcal{V}$ satisfies no nontrivial congruence identity, then it contains some algebra $\mathbf{A}$ that has a nontrivial rectangular tolerance T. If $\mathcal{V}$ has a join term, then it acts as a compatible semilattice operation on T. Can this fact be used to prove that the congruence lattice of an arbitrary semilattice is embeddable into the congruence lattice of some subdirect power of $\mathbf{A}$? (This would give an affirmative answer to the second part of Problem P1.) If $\mathcal{V}$ does not have a join term, will a similar argument work using compatible relations in place of the join term?

If the answer to the first part of Problem P1 is affirmative, then the congruence prevariety of the variety of semilattices is a unique minimal congruence prevariety that generates the variety of all lattices. This leads to the next question.

PROBLEM P2. Determine all minimal congruence prevarieties.

Could it be that every minimal congruence prevariety is modular or equal to the congruence prevariety of the variety of semilattices?

PROBLEM P3. Are there only countably many join semidistributive congruence varieties?

It is proved in Theorem 9 of [**36**] that there are continuum many different congruence varieties contained in the congruence variety of abelian groups. Problem P3 asks whether there are relatively few congruence varieties of varieties omitting abelian congruences.

PROBLEM P4. Let $\mathbf{A}$ be an arbitrary algebra. Suppose $\mathbf{Con}(\mathbf{A})$ contains congruences such that $\alpha \vee \beta = \theta$, $\alpha \vee \gamma = \theta$ and $\beta \wedge \gamma = 0$. Does the interval $I[\alpha, \theta]$ contain some instance of abelianness or rectangulation? (Specifically, must some quotient $\mathbf{A}/\delta$, $\alpha \leq \delta < \theta$ have a nontrivial abelian or rectangular tolerance?) If so, how can it be identified?

Lemma 5.19 of [**34**] proves that Problem P4 has an affirmative solution if $\mathbf{A}$ is finite, namely join semidistributivity failures involve nontrivial abelian or rectangular tolerances. Our Theorem 8.14 proves that if $\mathcal{V}$ has no algebra with a nontrivial abelian or rectangular tolerance, then it has no algebra with a join semidistributivity failure. Problem P4 asks whether our result can be localized to single algebras, as Hobby and McKenzie have done for finite algebras.

PROBLEM P5. Investigate rectangular nilpotence and solvability in varieties with a join term. Can anything interesting be discovered about rectangularly nilpotent or solvable varieties that have a join term?

In this monograph we have examined only the 'abelianness concept' associated to rectangulation. Problem P5 suggests the next step.

PROBLEM P6. Is it true that every idempotent variety that is not interpretable in the variety of distributive lattices is congruence n-permutable for some n?

Hobby and McKenzie have proved a number of type omitting theorems for locally finite varieties in Chapter 9 of [**34**]. An affirmative solution to Problem P6 will remove the remaining obstacle to extending all of these type omitting theorems to arbitrary varieties.

PROBLEM P7. Develop and apply a commutator theory for the weak term condition. Is there a Klukovits-type characterization of weakly abelian varieties?

The weak term condition has been investigated in [**44, 49, 50**]. Homomorphic images of finite Abelian algebras are weakly Abelian and every residually small variety generated by a finite nilpotent algebra is also weakly Abelian. A better understanding of this concept could be achieved by giving an equational-type characterization of weakly Abelian varieties related to Theorem 7.7 of [**49**].

APPENDIX A

Varieties with Special Terms

Throughout the text there are theorems characterizing the two classes of varieties

(i) possessing a Taylor term or
(ii) possessing a Hobby-McKenzie term.

These characterizations are collected here.

A.1. Varieties with a Taylor Term

THEOREM A.1. *A variety $\mathcal{V}$ has a Taylor term*[1] *if and only if one of the following conditions is satisfied.*

(1) *$\mathcal{V}$ satisfies an idempotent Maltsev condition that fails in some variety. (Thm. 2.14)*
(2) *$\mathcal{V}$ satisfies an idempotent Maltsev condition that fails in the variety of sets. (Thm. 2.14)*
(3) *$\mathcal{V}$ has no member with a nonzero strongly abelian tolerance. (Thm. 3.13)*
(4) *$\mathcal{V}$ has no member with a nonzero strongly abelian congruence. (Thm. 3.13)*
(5) *There exist $m \geq 1$, $n \geq 1$, idempotent sixary terms $f_1, \ldots, f_m, g_1, \ldots, g_n$ and a binary term s such that the following identities hold in $\mathcal{V}$:*
   (i) $s(x,y) \approx f_1(y,y,x,x,x,y)$,
   (ii) $f_m(y,x,x,x,y,y) \approx x$,
   (iii) $f_i(x,x,x,y,y,y) \approx f_i(y,y,x,x,x,y)$, $1 \leq i \leq m$,
   (iv) $f_i(y,x,x,x,y,y) \approx f_{i+1}(y,y,x,x,x,y)$, $1 \leq i \leq m-1$,
   (v) $s(x,y) \approx g_1(y,y,x,x,x,y)$,
   (vi) $g_n(y,x,x,x,y,y) \approx y$,
   (vii) $g_i(x,x,x,y,y,y) \approx g_i(y,y,x,x,x,y)$, $1 \leq i \leq n$,
   (viii) $g_i(y,x,x,x,y,y) \approx g_{i+1}(y,y,x,x,x,y)$, $1 \leq i \leq n-1$. *(Thm. 3.21)*
(6) *$\mathcal{V}$ has a join term*[2]. *(Thm. 3.21)*
(7) *$\mathcal{V}$ has no member with a nonzero strongly rectangular*[3] *tolerance. (Thm. 3.23)*
(8) *$\mathcal{V}$ has no member with a nonzero strongly rectangular congruence. (Thm. 3.23)*
(9) *The quasi-identity*

$$\big((p \wedge q) \approx s\big) \;\&\; \big((p \wedge r) \approx s\big) \;\&\; \big(p_{[2]} \approx s\big) \;\rightarrow\; \big((p \wedge (q \vee r)) \approx s\big)$$

holds in the congruence lattices of algebras in $\mathcal{V}$.[4] *(Thm. 4.12)*

---

[1]Definition 2.15.
[2]Definition 3.19.
[3]Definition 3.1.
[4]The lattice word $p_{[2]}$ is defined on page 59.

(10) *The quasi-identity of Item (9) holds in the meet continuous congruence variety of $\mathcal{V}$. (Thm. 4.12)*
(11) *$\mathcal{V}$ satisfies a nontrivial meet continuous congruence identity. (Thm. 4.12)*
(12) *$\mathcal{V}$ satisfies a congruence inclusion of the form*

$$\alpha \cap (\beta \circ_4 \gamma) \subseteq w(\alpha, \beta, \gamma)$$

for some lattice word $w(p,q,r)$ such that $w(p,q,r) < p \wedge (q \vee r)$ in the free lattice $\mathbf{F}_{\mathcal{L}}(\{p,q,r\})$. (Thm. 4.12)
(13) *For some $k \geq 4$, $\mathcal{V}$ satisfies a congruence inclusion of the form*

$$\alpha \cap (\beta \circ_k \gamma) \subseteq w(\alpha, \beta, \gamma),$$

where $w(p,q,r) < p \wedge (q \vee r)$ in $\mathbf{F}_{\mathcal{L}}(\{p,q,r\})$. (Thm. 4.12)
(14) *$\mathcal{V}$ satisfies a congruence inclusion of the form*

$$\alpha \cap (\beta \circ_3 \gamma) \subseteq w(\alpha, \beta, \gamma),$$

where $w(p,q,r)$ is a lattice word such that $w(p,q,r) \not\geq p^n$ for any n in the free lattice $\mathbf{F}_{\mathcal{L}}(\{p,q,r\})$.[5] (Proved in the remarks after Thm. 4.12)
(15) *For any finitely presented lattice **L** satisfying (W) but failing quasi-identity (4.4), **L** does not appear as a sublattice of **Con**(**A**) for any $\mathbf{A} \in \mathcal{V}$. (For example, **L** could be $\mathbf{D}_1$, $\mathbf{E}_1$ or **G**.[6]) (Thm. 4.23)*
(16) *For any finitely presented lattice **L** satisfying (W) but failing quasi-identity (4.4), **L** is not in the meet continuous congruence variety of $\mathcal{V}$. (Thm. 4.23)*
(17) *$\mathcal{V}$ has no member with a nonzero tolerance that is simultaneously rectangular[7] and abelian. (Cor. 5.15)*
(18) *$\mathcal{V}$ has no member with a nonzero congruence that is simultaneously rectangular and abelian. (Cor. 5.15)*

A.2. Varieties with a Hobby-McKenzie Term

THEOREM A.2. *A variety $\mathcal{V}$ has a Hobby-McKenzie term[8] if and only if one of the following conditions is satisfied.*

(1) *$\mathcal{V}$ satisfies an idempotent Maltsev condition that fails in the variety of semilattices. (Thm. 2.16)*
(2) *$\mathcal{V}$ has no member with a nontrivial rectangular[9] tolerance. (Thm. 5.25)*
(3) *$\mathcal{V}$ has no member with a nontrivial rectangular congruence. (Thm. 5.22)*
(4) *$x \sim_T y$ in $\mathbf{F} = \mathbf{F}_{\mathcal{V}}(x,y)$ for $T = \mathrm{Tg}^{\mathbf{F}}(x,y)$.[10] (Thm. 5.23)*
(5) *There exists an $m \geq 1$, and sixary terms $f_1, \ldots, f_m$ such that $\mathcal{V}$ satisfies the following identities:*
    (i) $x \approx f_1(y,y,x,x,x,y)$,
    (ii) $f_m(y,x,x,x,y,y) \approx y$,
    (iii) $f_i(x,x,x,y,y,y) \approx f_i(y,y,x,x,x,y)$, $1 \leq i \leq m$,
    (iv) $f_i(y,x,x,x,y,y) \approx f_{i+1}(y,y,x,x,x,y)$, $1 \leq i \leq m-1$. *(Thm. 5.23)*
(6) *For every algebra $\mathbf{A} \in \mathcal{V}$, tolerance S on **A**, and $(a,b) \in S$, $a \geq_S b$ holds. (Thm. 5.23)*

---

[5]The lattice word p^n is defined on page 63.
[6]These lattices are first defined in Theorem 2.2.
[7]Definitions 5.1 and 5.3.
[8]Definition 2.17.
[9]Definitions 5.1 and 5.3.
[10]The relations $\sim_T$ and $\geq_T$ are defined on page 40.

(7) *For every algebra* $\mathbf{A} \in \mathcal{V}$, *tolerance* S *on* $\mathbf{A}$, *and* $(a,b) \in S$, $a \sim_S b$ *holds.* *(Thm.* 5.23*)*
(8) *Whenever* θ, μ, ν *and* δ *are congruences on some* $\mathbf{A} \in \mathcal{V}$, *then* $(\theta \circ \mu) \cap (\nu \circ \delta) \subseteq [(\theta \vee \delta) \wedge (\nu \vee \mu)] \vee \delta \vee \mu$. *(Thm.* 5.28*)*
(9) $\mathcal{V}$ *has a sequence of terms* $f_i(x,y,u,v)$ *for* $0 \le i \le 2m+1$, *such that*
   (i) $\mathcal{V} \models f_0(x,y,u,v) \approx x$;
   (ii) $\mathcal{V} \models f_i(x,y,y,y) \approx f_{i+1}(x,y,y,y)$ *for even* i;
   (iii) $\mathcal{V} \models f_i(x,x,y,y) \approx f_{i+1}(x,x,y,y)$ *and*
   $\mathcal{V} \models f_i(x,y,x,y) \approx f_{i+1}(x,y,x,y)$ *for odd* i;
   (iv) $\mathcal{V} \models f_{2m+1}(x,y,u,v) \approx v$. *(Thm.* 5.28*)*
(10) $\mathcal{V}$ *omits special* $\mathbf{D}_2$*'s. (Thm.* 5.28*)*
(11) $\mathcal{V}$ *satisfies a nontrivial congruence identity. (Thm.* 7.15*)*
(12) *There exists* $\ell < \omega$ *such that the lattice* $\mathbf{N}_{\ell+5}$ *cannot be embedded into* $\mathbf{Con}(\mathbf{A})$ *for any* $\mathbf{A} \in \mathcal{V}$.[11] *(Thm.* 7.15*)*
(13) *There is a positive integer* N *such that no algebra in* $\mathcal{V}$ *has a herringbone of length greater than* N *in its congruence lattice. (Thm.* 8.3*)*
(14) $\mathcal{V}$ *satisfies the congruence identity*

$$z^M \approx z^{M+1}$$

for some M.[12] *(Thm.* 8.3*)*
(15) *For every* $\mathbf{A} \in \mathcal{V}$, $\mathbf{Con}(\mathbf{A})$ *has an* $\mathrm{SD}_\vee$/Modular *factorization in the category of meet continuous lattices. (Thm.* 8.10*)*
(16) *For every* $\mathbf{A} \in \mathcal{V}$, $\mathbf{Con}(\mathbf{A})$ *has an* $\mathrm{SD}_\vee$/Modular *factorization in the category of ordinary lattices. (Thm.* 8.10*)*
(17) *For every* $\mathbf{A} \in \mathcal{V}$, *each* $\mathrm{SD}_\vee$*-failure*[13] *in* $\mathbf{Con}(\mathbf{A})$ *is modular. (Thm.* 8.10*)*
(18) *If* $\mathbf{L}$ *belongs to* $\mathrm{CON}(\mathcal{V})$, *then each* $\mathrm{SD}_\vee$*-failure in* $\mathbf{L}$ *is modular.* *(Thm.* 8.10*)*
(19) $\mathbf{D}_2$ *does not appear as a sublattice of* $\mathbf{Con}(\mathbf{A})$ *for any* $\mathbf{A} \in \mathcal{V}$. *(Thm.* 8.11*)*
(20) $\mathbf{D}_2 \notin \mathrm{CON}(\mathcal{V})$. *(Thm.* 8.11*)*
(21) *The quasi-identity*

$$\big((p \vee q) \approx s\big) \;\&\; \big((p \vee r) \approx s\big) \;\&\; \big(p^{[2]} \approx s\big) \to \big((p \vee (q \wedge r)) \approx s\big)$$

holds in the congruence lattices of algebras in $\mathcal{V}$.[14] *(Thm.* 8.12*)*
(22) *The quasi-identity of Item (20) holds in* $\mathrm{CON}(\mathcal{V})$. *(Thm.* 8.12*)*
(23) $\mathcal{V} \models_{con} \alpha \cap (\beta \circ \gamma) \subseteq \big(\gamma \vee (\alpha \wedge \beta)\big) \circ \big(\beta \vee (\alpha \wedge \gamma)\big)$. *(Thm.* 8.13*)*
(24) *There is a positive integer* k *and ternary terms* $d_0, \dots, d_{2k+1}, e_0, \dots, e_{2k+1}$, p *such that* $\mathcal{V}$ *satisfies the following equations:*
   (i) $d_0(x,y,z) \approx p(x,y,z) \approx e_0(x,y,z)$;
   (ii) $d_i(x,y,y) \approx d_{i+1}(x,y,y)$ *and* $e_i(x,x,y) \approx e_{i+1}(x,x,y)$ *for even* i;
   (iii) $d_i(x,x,y) \approx d_{i+1}(x,x,y)$, $d_i(x,y,x) \approx d_{i+1}(x,y,x)$,
   $e_i(x,y,y) \approx e_{i+1}(x,y,y)$ *and* $e_i(x,y,x) \approx e_{i+1}(x,y,x)$ *for odd* i;
   (iv) $d_{2k+1}(x,y,z) \approx x$ *and* $e_{2k+1}(x,y,z) \approx z$. *(Thm.* 8.13*)*

---

[11]The lattice $\mathbf{N}_{\ell+5}$ is depicted in Figure 7.8 on page 128.
[12]The word z^M is defined on page 136.
[13]The concept of $\mathrm{SD}_\vee$-failure is defined on page 14.
[14]This quasi-identity is the dual of the one in Theorem A.1 (9).

Bibliography

[1] G. Birkhoff, *Lattice Theory*, Corrected reprint of the 1967 third edition, American Mathematical Society Colloquium Publications, **25**, American Mathematical Society, Providence, R.I., 1984. MR0227053 (37:2638)

[2] S. Burris and H. P. Sankappanavar, *A Course in Universal Algebra*, Graduate Texts in Mathematics No. 78, Springer–Verlag, 1981. MR648287 (83k:08001)

[3] G. Czédli, *A Mal'cev type condition for the semi-distributivity of congruence lattices*, Acta Sci. Math. **43** (1981), 267–272. MR640303 (83h:08012)

[4] G. Czédli, *A characterization for congruence semi-distributivity*, in *Universal Algebra and Lattice Theory*, Springer Lecture Notes in Mathematics **1004**, 1983, 104–110. MR716177 (85g:08006)

[5] G. Czédli, *Mal'cev conditions for Horn sentences with congruence permutability*, Acta Math. Hung. **44** (1984), 115–124. MR759039 (87a:08007)

[6] G. Czédli and A. Day, *Horn sentences with* (W) *and weak Mal'cev conditions*, Algebra Universalis **19** (1984), 217–230. MR758319 (87a:08006)

[7] B. A. Davey, W. Poguntke, and I. Rival, *A characterization of semi-distributivity*, Algebra Universalis **5** (1975), 72–75. MR0382103 (52:2991)

[8] A. Day, *A characterization of modularity for congruence lattices of algebras*, Canad. Math. Bull. **12** (1969), 167–173. MR0248063 (40:1317)

[9] A. Day, *p-Modularity implies modularity in equational classes*, Algebra Universalis **3** (1973), 398–399. MR0354497 (50:6975)

[10] A. Day, *Splitting lattices and congruence modularity*, *Contributions to Universal Algebra* (Colloq., József Attila Univ., Szeged, 1975), pp. 57–71, Colloq. Math. Soc. János Bolyai **17**, North-Holland, Amsterdam, 1977. MR0552770 (58:27694)

[11] A. Day and R. Freese, *A characterization of identities implying congruence modularity, I*, Canad. J. Math. **32** (1980), 1140–1167. MR596102 (82b:08009)

[12] A. Day and E. W. Kiss, *Frames and rings in congruence modular varieties*, J. Algebra **109** (1987), 479–507. MR902965 (89c:08007)

[13] R. Dedekind, *Über die von drei Moduln erzeugte Dualgruppe*, Math. Ann. **53** (1900), 371–403. MR1511094

[14] S. Fajtlowicz and J. Schmidt, *Bézout families, join-congruences, and meet-irreducible ideals*, *Lattice theory* (Proc. Colloq., Szeged, 1974), pp. 51–76. Colloq. Math. Soc. János Bolyai, **14**, North-Holland, Amsterdam, 1976. MR0444533 (56:2883)

[15] R. Freese, *Ideal lattices of lattices*, Pacific J. Math. **57** (1975), 125–133. MR0371751 (51:7968)

[16] R. Freese, J. Jezek, and J. B. Nation, *Free Lattices*, Mathematical Surveys and Monographs **42**, American Math. Soc., 1995. MR1319815 (96c:06013)

[17] R. Freese and B. Jónsson, *Congruence modularity implies the Arguesian identity*, Algebra Universalis **6** (1976), 225–228. MR0472644 (57:12340)

[18] R. Freese, K. Kearnes and J. B. Nation, *Congruence lattices of congruence semidistributive algebras*, Lattice theory and its applications, 63–78, Res. Exp. Math. **23**, Heldermann, Lemgo, 1995. MR1366865 (97a:08015)

[19] R. Freese and R. McKenzie, *Commutator Theory for Congruence Modular Varieties*, LMS Lecture Notes **125**, Cambridge University Press, 1987. MR909290 (89c:08006)

[20] R. Freese and J. B. Nation, *Congruence lattices of semilattices*, Pacific J. Math. **49** (1973), 51–58. MR0332590 (48:10916)

[21] R. Freese and J. B. Nation, *3-3 lattice inclusions imply congruence modularity*, Algebra Universalis **7** (1977), 191–194. MR0434906 (55:7870)

[22] N. Funayama and T. Nakayama, *On the distributivity of a lattice of lattice congruences*, Proc. Imp. Acad. Tokyo **18** (1942), 553–554. MR0014065 (7:236c)

[23] C. F. Gauss, *Disquisitiones Arithmeticae*, Lipsiae, in commissis apud Gerh. Fleischer, Jun., 1801.

[24] O. C. García and W. Taylor, *The lattice of interpretability types of varieties*, Mem. Amer. Math. Soc. **50**, 1984. MR749524 (86e:08006a)

[25] V. A. Gorbunov, *Algebraic Theory of Quasivarieties*, Translated from the Russian. Siberian School of Algebra and Logic. Consultants Bureau, New York, 1998. MR1654844 (2001a:08004)

[26] G. Grätzer and E. T. Schmidt, *Characterizations of congruence lattices of abstract algebras*, Acta Sci. Math. **24** (1963), 34–59. MR0151406 (27:1391)

[27] H. P. Gumm, *Geometrical Methods in Congruence Modular Algebras*, Mem. Amer. Math. Soc. **45**, 1983. MR714648 (85e:08012)

[28] J. Hagemann and C. Herrmann, *A concrete ideal multiplication for algebraic systems and its relation to congruence distributivity*, Arch. Math. (Basel) **32** (1979), 234–245. MR541622 (80j:08006)

[29] J. Hagemann and A. Mitschke, *On n-permutable congruences*, Algebra Universalis **3** (1973), 8–12. MR0330010 (48:8349)

[30] M. Haiman, *Arguesian lattices which are not linear*, Bull. Amer. Math. Soc. (N.S.) **16** (1987), no. 1, 121–123. MR866029 (87m:06014)

[31] C. Herrmann, *Affine algebras in congruence modular varieties*, Acta Sci. Math. **41** (1979), 119–125. MR534504 (80h:08011)

[32] C. Herrmann and W. Poguntke, *Axiomatic classes of modules*, Darmstadt Preprint **12**, 1972.

[33] C. Herrmann and W. Poguntke, *The class of sublattices of normal subgroup lattices is not elementary*, Algebra Universalis **4** (1974), 280–286. MR0354458 (50:6936)

[34] D. Hobby and R. McKenzie, *The Structure of Finite Algebras*, Contemporary Mathematics v. 76, American Mathematical Society, 1988. MR958685 (89m:08001)

[35] W. Hodges, *Model Theory*, Encyclopedia of Mathematics and its Applications **42**, Cambridge University Press, Cambridge, 1993. MR1221741 (94e:03002)

[36] G. Hutchinson and G. Czédli, *A test for identities satisfied in lattices of submodules*, Algebra Universalis **8** (1978), 269–309. MR0469840 (57:9621)

[37] B. Jónsson, *Modular lattices and Desargues' Theorem*, Math. Scand. **2** (1954), 295–314. MR0067859 (16:787f)

[38] B. Jónsson, *Algebras whose congruence lattices are distributive*, Math. Scand. **21** (1967), 110–121. MR0237402 (38:5689)

[39] B. Jónsson, *Identities in congruence varieties*, Lattice theory (Proc. Colloq., Szeged, 1974), pp. 195–205, Colloq. Math. Soc. János Bolyai, **14**, North-Holland, Amsterdam, 1976. MR0439713 (55:12599)

[40] B. Jónsson, *Congruence varieties*, Algebra Universalis **10** (1980), 355–394. MR564122 (81e:08004)

[41] B. Jónsson and I. Rival, *Lattice varieties covering the smallest nonmodular variety*, Pacific J. Math. **82** (1979), 463–478. MR551703 (81j:06007)

[42] K. Kaarli and R. McKenzie, *Affine complete varieties are congruence distributive*, Algebra Universalis **38** (1997), 329–354. MR1620035 (99m:08010)

[43] K. Kaarli and A. Pixley, *Affine complete varieties*, Algebra Universalis **24** (1987), 74–90. MR921532 (88k:08002)

[44] K. A. Kearnes, *An order-theoretic property of the commutator*, Internat. J. Algebra Comput. **3** (1993), 491–533. MR1250248 (95c:08002)

[45] K. A. Kearnes, *Varieties with a difference term*, J. Algebra **177** (1995), 926–960. MR1358491 (97c:08007)

[46] K. A. Kearnes, *A characterization of locally finite varieties that satisfy a nontrivial congruence identity*, Algebra Universalis **42** (1999), 195–204. MR1736714 (2001d:08012)

[47] K. A. Kearnes, *Almost all minimal idempotent varieties are congruence modular*, Algebra Universalis **44** (2000), 39–45. MR1801632 (2001j:08008)

[48] K. A. Kearnes, *Congruence join semidistributivity is equivalent to a congruence identity*, Algebra Universalis **46** (2001), 373–387. MR1857204 (2002f:08008)

[49] K. A. Kearnes and E. W. Kiss, *Finite algebras of finite complexity*, Discrete Math. **207** (1999), 89–135. MR1710485 (2000k:08001)

[50] K. A. Kearnes and E. W. Kiss, *Residual smallness and weak centrality*, International Journal of Algebra and Computation **13** (2003), 35–59. MR1970866 (2004c:08012)

[51] K. A. Kearnes and E. W. Kiss, *The triangular principle is equivalent to the triangular scheme*, Algebra Universalis **54** (2005), no. 3, 373–383. MR2219417 (2007d:08011)

[52] K. A. Kearnes and J. B. Nation, *Axiomatizable and nonaxiomatizable congruence prevarieties*, Algebra Universalis **59** (2008), no. 3-4, 323–335. MR2470584 (2009k:08004)

[53] K. A. Kearnes and Á. Szendrei, *The relationship between two commutators*, Internat. J. Algebra Comput. **8** (1998), 497–531. MR1663558 (2000e:08001)

[54] K. A. Kearnes and M. A. Valeriote, *A modification of Polin's variety*, Algebra Universalis **41** (1999), 229–231. MR1699342 (2000d:08004)

[55] E. W. Kiss and P. Pröhle, *Problems and results in tame congruence theory*, Algebra Universalis **29** (1992), 151–171. MR1157431 (93g:08004)

[56] P. Lipparini, *Congruence identities satisfied in n-permutable varieties*, Boll. Un. Mat. Ital. B (7) **8** (1994), no. 4, 851–868. MR1315822 (95m:08006)

[57] P. Lipparini, *Commutator theory without join-distributivity*, Trans. Amer. Math. Soc. **346** (1994), 177–202. MR1257643 (95c:08009)

[58] P. Lipparini, *n-permutable varieties satisfy nontrivial congruence identities*, Algebra Universalis **33** (1995), 159–168. MR1318980 (96c:08010)

[59] P. Lipparini, *A characterization of varieties with a difference term*, Canad. Math. Bull. **39** (1996), 308–315. MR1411074 (97h:08010)

[60] P. Lipparini, *A characterization of varieties with a difference term, II: neutral = meet semidistributive*, Canad. Math. Bull. **41** (1998), 318–327. MR1637665 (2000a:08021)

[61] P. Lipparini, *An elementary proof that n-permutable varieties satisfy lattice identities*, manuscript available at `http://www.mat.uniroma2.it/~lipparin/nperm2.dvi`

[62] P. Lipparini, *Every m-permutable variety satisfies the congruence identity* $\alpha\beta_h = \alpha\gamma_h$, Proc. Amer. Math. Soc. **136** (2008), 1137–1144. MR2367087 (2008i:08001)

[63] A. I. Maltsev, *On the general theory of algebraic systems* (in Russian), Mat. Sb. N. S. **35**(77) (1954), 3–20. MR0065533 (16:440e)

[64] S. Mac Lane, *Categories for the Working Mathematician*, Second edition, Graduate Texts in Mathematics **5**, Springer–Verlag, 1998. MR1712872 (2001j:18001)

[65] R. McKenzie, *Equational bases and nonmodular lattice varieties*, Trans. Amer. Math. Soc. **174** (1972), 1–43. MR0313141 (47:1696)

[66] R. McKenzie, *Some unsolved problems between lattice theory and equational logic*, Proceedings of the University of Houston Lattice Theory Conference (Houston, Tex., 1973), pp. 564–573, Dept. Math., Univ. Houston, Houston, Tex., 1973. MR0398920 (53:2771)

[67] R. McKenzie, *Finite forbidden lattices*, in *Universal Algebra and Lattice Theory*, Lecture Notes in Mathematics **1004**, Springer-Verlag, 1983. MR716183 (85b:06006)

[68] R. McKenzie, G. McNulty, and W. Taylor, *Algebras, Lattices, Varieties, Volume I*, The Wadsworth & Brooks/Cole Mathematics Series. Wadsworth & Brooks/Cole Advanced Books & Software, Monterey, CA, 1987. MR883644 (88e:08001)

[69] P. Mederly, *Three Mal'cev type theorems and their application*, Mat. Časopis Sloven. Akad. Vied **25** (1975), 83–95. MR0384650 (52:5523)

[70] J. B. Nation, *Varieties whose congruences satisfy certain lattice identities*, Algebra Universalis **4** (1974), 78–88. MR0354501 (50:6979)

[71] N. Newrly and M. A. Valeriote, *Some generalizations of Polin's variety*, manuscript, 1999.

[72] P. P. Pálfy and Cs. Szabó, *Congruence varieties of groups and abelian groups*, Lattice theory and its applications (Darmstadt, 1991), 163–183, Res. Exp. Math. **23** Heldermann, Lemgo, 1995. MR1366871 (96k:20052)

[73] A. Pixley, *Distributivity and permutability of congruence relations in equational classes of algebras*, Proc. Amer. Math. Soc. **14** (1963), 105–109. MR0146104 (26:3630)

[74] A. Pixley, *Local Mal'cev conditions*, Canad. Math. Bull. **15** (1972), 559–568. MR0309837 (46:8942)

[75] S. V. Polin, *Identities in congruence lattices of universal algebras*, Mat. Zametki **22** (1977), 443–451. MR0491407 (58:10663)

[76] J. D. H. Smith, *Mal'cev Varieties*, Lecture Notes in Mathematics **554**, Springer-Verlag, 1976. MR0432511 (55:5499)

[77] W. Taylor, *Characterizing Mal'cev conditions*, Algebra Universalis **3** (1973), 351–397. MR0349537 (50:2030)

[78] W. Taylor, *Varieties obeying homotopy laws*, Canadian J. Math. **29** (1977), 498–527. MR0434928 (55:7891)

[79] R. Wille, *Kongruenzklassengeometrien*, Lecture Notes in Mathematics **113**, Springer-Verlag, 1970. MR0262149 (41:6759)

Index

Editorial Information

To be published in the *Memoirs*, a paper must be correct, new, nontrivial, and significant. Further, it must be well written and of interest to a substantial number of mathematicians. Piecemeal results, such as an inconclusive step toward an unproved major theorem or a minor variation on a known result, are in general not acceptable for publication.

Papers appearing in *Memoirs* are generally at least 80 and not more than 200 published pages in length. Papers less than 80 or more than 200 published pages require the approval of the Managing Editor of the Transactions/Memoirs Editorial Board. Published pages are the same size as those generated in the style files provided for $\mathcal{A}\mathcal{M}\mathcal{S}$-LaTeX or $\mathcal{A}\mathcal{M}\mathcal{S}$-TeX.

Information on the backlog for this journal can be found on the AMS website starting from `http://www.ams.org/memo`.

A Consent to Publish is required before we can begin processing your paper. After a paper is accepted for publication, the Providence office will send a Consent to Publish and Copyright Agreement to all authors of the paper. By submitting a paper to the *Memoirs*, authors certify that the results have not been submitted to nor are they under consideration for publication by another journal, conference proceedings, or similar publication.

Information for Authors

Memoirs is an author-prepared publication. Once formatted for print and on-line publication, articles will be published as is with the addition of AMS-prepared frontmatter and backmatter. Articles are not copyedited; however, confirmation copy will be sent to the authors.

Initial submission. The AMS uses Centralized Manuscript Processing for initial submissions. Authors should submit a PDF file using the Initial Manuscript Submission form found at `www.ams.org/submission/memo`, or send one copy of the manuscript to the following address: Centralized Manuscript Processing, MEMOIRS OF THE AMS, 201 Charles Street, Providence, RI 02904-2294 USA. If a paper copy is being forwarded to the AMS, indicate that it is for *Memoirs* and include the name of the corresponding author, contact information such as email address or mailing address, and the name of an appropriate Editor to review the paper (see the list of Editors below).

The paper must contain a *descriptive title* and an *abstract* that summarizes the article in language suitable for workers in the general field (algebra, analysis, etc.). The *descriptive title* should be short, but informative; useless or vague phrases such as "some remarks about" or "concerning" should be avoided. The *abstract* should be at least one complete sentence, and at most 300 words. Included with the footnotes to the paper should be the 2010 *Mathematics Subject Classification* representing the primary and secondary subjects of the article. The classifications are accessible from `www.ams.org/msc/`. The Mathematics Subject Classification footnote may be followed by a list of *key words and phrases* describing the subject matter of the article and taken from it. Journal abbreviations used in bibliographies are listed in the latest *Mathematical Reviews* annual index. The series abbreviations are also accessible from `www.ams.org/msnhtml/serials.pdf`. To help in preparing and verifying references, the AMS offers MR Lookup, a Reference Tool for Linking, at `www.ams.org/mrlookup/`.

Electronically prepared manuscripts. The AMS encourages electronically prepared manuscripts, with a strong preference for $\mathcal{A}\mathcal{M}\mathcal{S}$-LaTeX. To this end, the Society has prepared $\mathcal{A}\mathcal{M}\mathcal{S}$-LaTeX author packages for each AMS publication. Author packages include instructions for preparing electronic manuscripts, samples, and a style file that generates the particular design specifications of that publication series. Though $\mathcal{A}\mathcal{M}\mathcal{S}$-LaTeX is the highly preferred format of TeX, author packages are also available in $\mathcal{A}\mathcal{M}\mathcal{S}$-TeX.

Authors may retrieve an author package for *Memoirs of the AMS* from `www.ams.org/journals/memo/memoauthorpac.html` or via FTP to `ftp.ams.org` (login as `anonymous`, enter your complete email address as password, and type `cd pub/author-info`). The

AMS Author Handbook and the *Instruction Manual* are available in PDF format from the author package link. The author package can also be obtained free of charge by sending email to tech-support@ams.org or from the Publication Division, American Mathematical Society, 201 Charles St., Providence, RI 02904-2294, USA. When requesting an author package, please specify $\mathcal{A}_{\mathcal{M}}\mathcal{S}$-LaTeX or $\mathcal{A}_{\mathcal{M}}\mathcal{S}$-TeX and the publication in which your paper will appear. Please be sure to include your complete mailing address.

After acceptance. The source files for the final version of the electronic manuscript should be sent to the Providence office immediately after the paper has been accepted for publication. The author should also submit a PDF of the final version of the paper to the editor, who will forward a copy to the Providence office.

Accepted electronically prepared files can be submitted via the web at www.ams.org/submit-book-journal/, sent via FTP, or sent on CD to the Electronic Prepress Department, American Mathematical Society, 201 Charles Street, Providence, RI 02904-2294 USA. TeX source files and graphic files can be transferred over the Internet by FTP to the Internet node ftp.ams.org (130.44.1.100). When sending a manuscript electronically via CD, please be sure to include a message indicating that the paper is for the *Memoirs*.

Electronic graphics. Comprehensive instructions on preparing graphics are available at www.ams.org/authors/journals.html. A few of the major requirements are given here.

Submit files for graphics as EPS (Encapsulated PostScript) files. This includes graphics originated via a graphics application as well as scanned photographs or other computer-generated images. If this is not possible, TIFF files are acceptable as long as they can be opened in Adobe Photoshop or Illustrator.

Authors using graphics packages for the creation of electronic art should also avoid the use of any lines thinner than 0.5 points in width. Many graphics packages allow the user to specify a "hairline" for a very thin line. Hairlines often look acceptable when proofed on a typical laser printer. However, when produced on a high-resolution laser imagesetter, hairlines become nearly invisible and will be lost entirely in the final printing process.

Screens should be set to values between 15% and 85%. Screens which fall outside of this range are too light or too dark to print correctly. Variations of screens within a graphic should be no less than 10%.

Inquiries. Any inquiries concerning a paper that has been accepted for publication should be sent to memo-query@ams.org or directly to the Electronic Prepress Department, American Mathematical Society, 201 Charles St., Providence, RI 02904-2294 USA.

Editors

This journal is designed particularly for long research papers, normally at least 80 pages in length, and groups of cognate papers in pure and applied mathematics. Papers intended for publication in the *Memoirs* should be addressed to one of the following editors. The AMS uses Centralized Manuscript Processing for initial submissions to AMS journals. Authors should follow instructions listed on the Initial Submission page found at www.ams.org/memo/memosubmit.html.

Algebra, to ALEXANDER KLESHCHEV, Department of Mathematics, University of Oregon, Eugene, OR 97403-1222; e-mail: klesh@uoregon.edu

Algebraic geometry, to DAN ABRAMOVICH, Department of Mathematics, Brown University, Box 1917, Providence, RI 02912; e-mail: amsedit@math.brown.edu

Algebraic topology, to ALEJANDRO ADEM, Department of Mathematics, University of British Columbia, Room 121, 1984 Mathematics Road, Vancouver, British Columbia, Canada V6T 1Z2; e-mail: adem@math.ubc.ca

Algebraic topology, to SOREN GALATIUS, Department of Mathematics, Stanford University, Stanford, CA 94305; e-mail: transactions@lists.stanford.edu

Arithmetic geometry, to TED CHINBURG, Department of Mathematics, University of Pennsylvania, Philadelphia, PA 19104-6395; e-mail: math-tams@math.upenn.edu

Automorphic forms, representation theory and combinatorics, to DANIEL BUMP, Department of Mathematics, Stanford University, Building 380, Sloan Hall, Stanford, California 94305; e-mail: bump@math.stanford.edu

Combinatorics, to JOHN R. STEMBRIDGE, Department of Mathematics, University of Michigan, Ann Arbor, Michigan 48109-1109; e-mail: JRS@umich.edu

Commutative and homological algebra, to LUCHEZAR L. AVRAMOV, Department of Mathematics, University of Nebraska, Lincoln, NE 68588-0130; e-mail: avramov@math.unl.edu

Complex analysis and harmonic analysis, to MALABIKA PRAMANIK, Department of Mathematics, 1984 Mathematics Road, University of British Columbia, Vancouver, BC, Canada V6T 1Z2; e-mail: malabika@math.ubc.ca

Differential geometry and global analysis, to CHRIS WOODWARD, Department of Mathematics, Rutgers University, 110 Frelinghuysen Road, Piscataway, NJ 08854; e-mail: ctw@math.rutgers.edu

Dynamical systems and ergodic theory and complex analysis, to YUNPING JIANG, Department of Mathematics, CUNY Queens College and Graduate Center, 65-30 Kissena Blvd., Flushing, NY 11367; e-mail: Yunping.Jiang@qc.cuny.edu

Functional analysis and operator algebras, to NATHANIEL BROWN, Department of Mathematics, 320 McAllister Building, Penn State University, University Park, PA 16802; e-mail: nbrown@math.psu.edu

Geometric analysis, to WILLIAM P. MINICOZZI II, Department of Mathematics, Johns Hopkins University, 3400 N. Charles St., Baltimore, MD 21218; e-mail: trans@math.jhu.edu

Geometric topology, to MARK FEIGHN, Math Department, Rutgers University, Newark, NJ 07102; e-mail: feighn@andromeda.rutgers.edu

Harmonic analysis, representation theory, and Lie theory, to E. P. VAN DEN BAN, Department of Mathematics, Utrecht University, P.O. Box 80 010, 3508 TA Utrecht, The Netherlands; e-mail: E.P.vandenBan@uu.nl

Logic, to ANTONIO MONTALBAN, Department of Mathematics, The University of California, Berkeley, Evans Hall #3840, Berkeley, California, CA 94720; e-mail: antonio@math.berkeley.edu

Number theory, to SHANKAR SEN, Department of Mathematics, 505 Malott Hall, Cornell University, Ithaca, NY 14853; e-mail: ss70@cornell.edu

Partial differential equations, to GUSTAVO PONCE, Department of Mathematics, South Hall, Room 6607, University of California, Santa Barbara, CA 93106; e-mail: ponce@math.ucsb.edu

Partial differential equations and functional analysis, to ALEXANDER KISELEV, Department of Mathematics, University of Wisconsin-Madison, 480 Lincoln Dr., Madison, WI 53706; e-mail: kiselev@math.wisc.edu

Probability and statistics, to PATRICK FITZSIMMONS, Department of Mathematics, University of California, San Diego, 9500 Gilman Drive, La Jolla, CA 92093-0112; e-mail: pfitzsim@math.ucsd.edu

Real analysis and partial differential equations, to WILHELM SCHLAG, Department of Mathematics, The University of Chicago, 5734 South University Avenue, Chicago, IL 60615; e-mail: schlag@math.uchicago.edu

All other communications to the editors, should be addressed to the Managing Editor, ROBERT GURALNICK, Department of Mathematics, University of Southern California, Los Angeles, CA 90089-1113; e-mail: guralnic@math.usc.edu.

Selected Published Titles in This Series

1036 **Matthias Lesch, Henri Moscovici, and Markus J. Pflaum,** Connes-Chern Character for Manifolds with Boundary and Eta Cochains, 2012

1035 **Igor Burban and Bernd Kreussler,** Vector Bundles on Degenerations of Elliptic Curves and Yang-Baxter Equations, 2012

1034 **Alexander Kleshchev and Vladimir Shchigolev,** Modular Branching Rules for Projective Representations of Symmetric Groups and Lowering Operators for the Supergroup $Q(n)$, 2012

1033 **Daniel Allcock,** The Reflective Lorentzian Lattices of Rank 3, 2012

1032 **John C. Baez, Aristide Baratin, Laurent Freidel, and Derek K. Wise,** Infinite-Dimensional Representations of 2-Groups, 2012

1031 **Idrisse Khemar,** Elliptic Integrable Systems: A Comprehensive Geometric Interpretation, 2012

1030 **Ernst Heintze and Christian Groß,** Finite Order Automorphisms and Real Forms of Affine Kac-Moody Algebras in the Smooth and Algebraic Category, 2012

1029 **Mikhail Khovanov, Aaron D. Lauda, Marco Mackaay, and Marko Stošić,** Extended Graphical Calculus for Categorified Quantum sl(2), 2012

1028 **Yorck Sommerhäuser and Yongchang Zhu,** Hopf Algebras and Congruence Subgroups, 2012

1027 **Olivier Druet, Frédéric Robert, and Juncheng Wei,** The Lin-Ni's Problem for Mean Convex Domains, 2012

1026 **Mark Behrens,** The Goodwillie Tower and the EHP Sequence, 2012

1025 **Joel Smoller and Blake Temple,** General Relativistic Self-Similar Waves that Induce an Anomalous Acceleration into the Standard Model of Cosmology, 2012

1024 **Mats Boij, Juan C. Migliore, Rosa M. Miró-Roig, Uwe Nagel, and Fabrizio Zanello,** On the Shape of a Pure O-Sequence, 2012

1023 **Tadeusz Iwaniec and Jani Onninen,** n-Harmonic Mappings between Annuli, 2012

1022 **Maurice Duits, Arno B.J. Kuijlaars, and Man Yue Mo,** The Hermitian Two Matrix Model with an Even Quartic Potential, 2012

1021 **Arnaud Deruelle, Katura Miyazaki, and Kimihiko Motegi,** Networking Seifert Surgeries on Knots, 2012

1020 **Dominic Joyce and Yinan Song,** A Theory of Generalized Donaldson-Thomas Invariants, 2012

1019 **Abdelhamid Meziani,** On First and Second Order Planar Elliptic Equations with Degeneracies, 2012

1018 **Nicola Gigli,** Second Order Analysis on $(\mathcal{P}_2(M), W_2)$, 2012

1017 **Zenon Jan Jabłoński, Il Bong Jung, and Jan Stochel,** Weighted Shifts on Directed Trees, 2012

1016 **Christophe Breuil and Vytautas Paškūnas,** Towards a Modulo p Langlands Correspondence for GL_2, 2012

1015 **Jun Kigami,** Resistance Forms, Quasisymmetric Maps and Heat Kernel Estimates, 2012

1014 **R. Fioresi and F. Gavarini,** Chevalley Supergroups, 2011

1013 **Kaoru Hiraga and Hiroshi Saito,** On L-Packets for Inner Forms of SL_n, 2011

1012 **Guy David and Tatiana Toro,** Reifenberg Parameterizations for Sets with Holes, 2011

1011 **Nathan Broomhead,** Dimer Models and Calabi-Yau Algebras, 2011

1010 **Greg Kuperberg and Nik Weaver,** A von Neumann Algebra Approach to Quantum Metrics/Quantum Relations, 2011

1009 **Tarmo Järvilehto,** Jumping Numbers of a Simple Complete Ideal in a Two-Dimensional Regular Local Ring, 2011

For a complete list of titles in this series, visit the AMS Bookstore at **www.ams.org/bookstore/memoseries/**.